Biological and Bio-Inspired Fluid Dynamics

David E. Rival

Biological and Bio-Inspired Fluid Dynamics

Theory and Application

 Springer

David E. Rival
Department of Mechanical and Materials
Engineering
Queen's University
Kingston, ON, Canada

ISBN 978-3-030-90273-5   ISBN 978-3-030-90271-1   (eBook)
https://doi.org/10.1007/978-3-030-90271-1

This Springer imprint is published by the registered company Springer Nature Switzerland AG
The registered company address is: Gewerbestrasse 11, 6330 Cham, Switzerland

To Laurissa, for her unwavering love and patience. And to our beautiful children, Audrey, Astrid, and Julien. Their boundless curiosity helps to remind us every day that all children are little researchers and scientists at heart.

Preface

This book serves as an introduction—or perhaps better yet—a *stepping stone*, between the very fundamentals of fluid dynamics and the rich, exciting, and often poorly understood world of biological, biomedical, and bio-inspired flows. As for the destination, the goal here is to solidify the basic tools through examples and to promote independent and creative problem-solving skills, such that the reader can later venture out into uncharted waters on their own. Inevitably for some, the material will at times read as a review. However, redundancy in the form of re-learning a concept from a different perspective can help reinforce one's understanding and promote new questions. For others, there will undoubtedly be gaps, but when paired with an introductory fluid-dynamics textbook and the willingness to tackle complex biological systems using *abstraction*, I believe the reader ultimately will flourish. Finally, this book deliberately avoids citing countless detailed scientific studies, be it analytical, numerical, or experimental, which appear daily and often displace the past studies with limited computing power or experimental tools. I hope that this approach will not offend my colleagues! Although at first a long literature review could seem very helpful to the reader, there is a risk that this approach might in fact dilute the fundamentals all the while distracting from the very goals of this book. However, it is assumed that the reader, once interested in a particular topic, will immediately delve into the literature and immerse themselves with the very latest and greatest. If this step takes place, then this book will have very much served its purpose.

Munich, Germany
July 2021

David E. Rival

Acknowledgments

This book would never have come together without all the many thoughtful contributions, during both course delivery and writing, from my talented graduate students and postdocs, including Josh Galler, Sonja Pejcic, Dr. Frieder Kaiser, Dr. Reza Najjari, Dr. Wenchao Yang, and Dr. Kai Zhang. I would especially like to thank two former PhD students, Dr. Mark Jeronimo and Dr. Adnan El Makdah, who not only worked tirelessly on putting material together but also offered countless suggestions to improve the content for a range of readers. Like any worthwhile endeavor, the journey was tortuous yet exciting at every twist and turn. From the initial support and encouragement to develop a course on the topic to the endless discussions with colleagues and friends from our ad hoc Bio-propulsion of Adaptive Systems group, the Queen's University Biological Station, and participants at our workshops and conferences from all corners of the world, I have been very fortunate for your comradery and willingness to cross existing boundaries. In particular, I must thank Dr. Gianluigi Bisleri (Toronto), Dr. Fran Bonier (Queen's), Dr. Christopher Cameron (Montreal), Dr. Jean-Bernard Caron (Royal Ontario Museum), Dr. John Dabiri (Caltech), Dr. Alex Dececchi (Mount Marty), Dr. Josephine Galipon (Keio), Dr. Melissa Green (Minnesota), Dr. Amer Johri (Queen's), Dr. Daegyoum Kim (KAIST), Dr. Jochen Kriegseis (Karlsruhe), Dr. Hans Larsson (McGill), Dr. Karen Mulleners (EPFL), Dr. Mike Rainbow (Queen's), Dr. Em Standen (Ottawa), Dr. Michael Triantafyllou (MIT), Dr. Andrew Walker (Calgary), Dr. Gabe Weymouth (Southampton), and the many other collaborators and friends for the inspiring conversations and feedback—it has been a beautiful journey and I look forward to many more trips! Last but not least, I must acknowledge both the Alexander von Humboldt Stiftung and Dr. Christian Kähler (Munich) who supported my sabbatical during which time this book came to fruition.

Contents

Chapter 1
Introduction

We start by exploring various perspectives (as well as misconceptions) when relating biology and fluid dynamics together. On the one hand, we have been inspired by nature for millennia. On the other hand, our insight into evolution and its nuances is quite recent in comparison. In fact, fluid dynamics and turbulence, albeit relatively *modern* fields of research, have varied far less in our basic understandings than, say, the rapidly evolving insights drawn from paleontology and genomics (e.g., epigenetics). Therefore, in this first chapter, we allow ourselves an opportunity to first consider the big picture before delving into the fundamental tools needed for future analysis. With this approach, we will be best prepared to revisit these more philosophical questions in later chapters, thus providing ourselves with the necessary framework with which to establish our own journeys for exploration.

1.1 Human-Centric Perspective

From the very beginning, floating *in utero*, we remain in constant observation—and awe!—of our complex yet elegant fluid surroundings. Since we inhabit and therefore perceive scales on the order of our bodies, it is no wonder that we tend to fixate on spatial and temporal scales that are easily relatable to us as individuals. Therefore, it should come as no surprise when, as a species, we struggle to make sense of the *invisible*, both of the tiny microscopic and huge geological scales that very much affect us at every level.

Furthermore, as we are social and compassionate beings and often are encouraged from a young age to focus on our anthropogenic activities, it should come as no surprise that we tend to establish a human-centric perspective early on. Despite every child's wildest fantasies about dinosaurs, the mention of a heart typically conjures up discussion of the *human* heart's four chambers rather than, for instance, unique biological solutions found in the invertebrate world (e.g., octopodes), or even

D. E. Rival, *Biological and Bio-Inspired Fluid Dynamics*,
https://doi.org/10.1007/978-3-030-90271-1_1

the incredible scales and adaptability to depth present in our blue-whale cousins (note their hearts are on the scale of a small car).

Although there is no harm in spinning our study of fluid dynamics through the lens of *biomedical* or *bioengineering* perspectives—in fact, this book often reverts back to the human body—it is important that at all steps along our journey we take the time to compare and contrast to other scales and *solutions* found in the vast biological world that surrounds us, be it from the present or even from the past. This comparative approach, with its roots in the classical study of biological systems, gives us perspective and allows us to pose unique questions. Through these comparisons (e.g., human versus blue-whale hearts), we can find the commonalities as well as the differences, and through that lens, we can learn a great deal more than when studying each system separately. It is this very act of removing oneself from the immediate problem/design at hand, and taking the time to step back and explore alternative solutions that can often help us tackle and solve the most challenging of problems. In other words, even when motivated by pressing biomedical challenges, there is merit not only in comparing to other *unique* solutions found today but in asking ourselves how our bodies resulted in their unique (and likely suboptimal) form.

In that regard, all we need to do is look at the profound discoveries of Jan Swammerdam (circa 1658, and published posthumously in Swammerdam (1950)), who has been credited with many key *medical* contributions such as the first observations of red blood cells, the various stages of animal development, and even mechanistic insights into constant-volume measurements of muscle contraction; see Fig. 1.1. It turns out that Swammerdam's work was not specifically focused on addressing human ailments but rather in comparing and contrasting the features of biological life as a whole. In fact, Swammerdam's name is associated with one

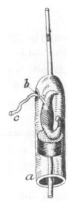

Fig. 1.1 Jan Swammerdam (left) whose influential works were among the first to document insect life cycles (centre), describe red blood cells, and measure the volume of muscles during active contraction (right)

particularly meaningful quote that we should all consider when embarking on this journey:

"The body of a beast deserves as great admiration as the human body."

For this reason, when referring to the current book's focus on *biofluids*, we must pay careful attention not to be distracted. For instance, *bio* of course refers to *biological* in this context, which we take as the study of nature's solutions constrained through evolution *and* by using a comparative approach. In a similar vein, *fluids* here refers to much more than the mere classification of various bulk properties found in nature, but really the study of *fluid dynamics*, where we develop and apply engineering analysis (i.e., mathematics, physics, and measurements) to characterize and solve complex fluids problems. Although undoubtedly each reader will have their specific motivations and interests, e.g., comparative biology, biomedical applications, biomimetics to develop new designs, or exploring the origins of life through the lens of paleontology, we are not going to take sides but rather let different problems inspire us to develop our own sets of unique questions.

1.2 Scaling in Biology and Fluid Dynamics

As we will see in later chapters, the concept of *scaling* is hugely important in the realm of fluid dynamics, and in particular, but not limited to the relentless exploration of turbulent flows. Therefore, it should come as no surprise that, in contrast to many other fields, fluid dynamicists rely heavily on non-dimensional groups to draw insight into a seemingly endless number of parameters. As with many fields of study, we can trace back to observations first documented by da Vinci (c. 1510). Here for instance in Fig. 1.2, we can marvel at da Vinci's sketches of various eddy (or vortex) length *scales* formed in a nearby canal. Not only did da Vinci observe the broad range of scales in fluid dynamics but he even imagined the soaring maneuvers birds would employ to adapt to flight in these various "turbulent" scales (see Fig. 1.2).

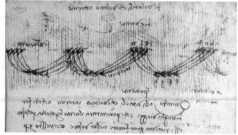

Fig. 1.2 Da Vinci sketched the various length scales of eddies while watching water flow into a nearby canal (left), as well as the dynamic soaring maneuvers of birds in turbulent winds (right)

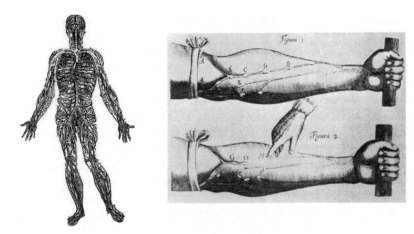

Fig. 1.3 Vesalius (1555) produces intricate sketches of the fractal distribution of the cardiovascular system (left), which were subsequently elaborated upon by Harvey (1628)

In contrast to our relentless search for scaling laws in fluid dynamics, the concept of scaling is fundamentally implicit in the biological world, be it the study of morphological change through development (i.e., allometry), or the evolution (and huge variation) of forms and scale across paleontological time, all the way to the fractal-like scaling of blood vessels present even in our very own "human-centric" cardiovascular system. The recurrence of shapes and patterns is shown quite elegantly through observations (and sketches) of the human cardiovascular system made by Vesalius (1555), and subsequent exploration by Harvey (1628); see Fig. 1.3. Here, the system is comprised of systematic steps down in scale starting from largest (the aorta carries blood away from the heart) through a series of branching arteries, and then broken down, in turn, further into arterioles, and finally at the near-microscopic scales of capillaries where oxygen is released. On the return voyage, the system scales back up from venules, to small veins and then finally to the largest veins returning to the heart. So naturally, we must ask ourselves whether small arterioles and venules are essentially the same as large arteries and veins just shrunken down in scale? What features are in common, and which ones are different? These are the kinds of questions that we face when exploring problems through the lens of scaling.

On that note, how does our human heart scale with other mammalian hearts like that of the aforementioned blue whale; see the relative size of the blue-whale aorta as a baby is inserted into the aorta in Fig. 1.4. Are all chambers and associated vessels more-or-less the same as that of an adult human heart, or if not, what has changed? And while we are on the topic of scaling, what happens to red blood cells in other mammals? Are blue-whale red blood cells a few orders of magnitude bigger as well, i.e., rather than say approximately 10 μm, as found in humans, are they now on the order of 1 mm instead? Indeed a strange question but when posed it does bring forth assumptions that we typically make.

Fig. 1.4 A blue-whale heart and its aorta compared to the scale of a baby begs the question: Is the rest of the whale's cardiovascular system, specifically blood constituents such as red blood cells, similarly scaled up?

1.3 Contingency vs. Evolutionary Convergence

At a philosophical level, we must step back and ask ourselves the following rhetorical question: Do we really believe that most examples found in nature are in fact *optimal* in their respective function? Unfortunately, this view is not atypical of engineering, or even medicine for that matter. To some extent, we find incredible parallels of *solutions* in nature, brought about through the process known as *evolutionary convergence*, which we discuss in detail later. On the other hand, evolution takes a complex path and we know that, not unlike the weather, there is a degree of randomness at play. This element of randomness or chaos is often referred to as *contingency*. For instance, take the asteroid that struck the Earth 66 million years ago (MYA) and effectively wiped out the stunning age of dinosaurs. Without that random event, would we humans (large mammals) be leading the Anthropocene today? To tackle these questions, let us first discuss how we perceive our biological world today and look at how far we have come in our understanding over the last centuries and even decades.

1.3.1 Organization and Biodiversity

It was not that long ago that we used to organize the world into simple categories such as inanimate objects (e.g., rocks), then plants, animals, and finally ourselves (humans) at the very top of the hierarchy. Thanks in part to our relentless need to organize, Linnaeus (c. 1735) developed the modern hierarchical classification system that we continue to modify to this day. Look at, for instance, *phylogenetic tree* by Haeckel (1866) describing our view of the living world in that period; see Fig. 1.5. Since then, we continue to move away from the human-centric perspective as we realize the enormous diversity present both at our visible scales and, even more so, under the microscope, as illustrated in Fig. 1.6. Compare our tiny little sliver among the Eukaryotes (*Opisthokonta*) within which the kingdoms of Animalia and Fungi thrive so inconspicuously in comparison to the huge array of diversity found among the Bacteria and Archaea.

1.3.2 Replaying Evolutionary History

Now that we have taken a humble step back from our human-centric view of diversity in the biological world, we recognize that perhaps our personal assumptions about biodiversity are limited to the scales within which we interact on a daily basis. It is now time to go back in time and restart evolutionary history, or as Stephen Jay Gould famously (yet contentiously) posed as a thought experiment in *Wonderful Life*, we should *replay the tape of life* (Gould 1990). Of course, the concept of *rewinding and replaying* dates to an era of cassette tapes, but the idea remains the same: What happens if we restart from the very beginning? Are humans or humanoids an inevitability, as suggested by Conway Morris (1999), or would possibly a big-brained dinosauroid, perhaps raptor-like, rule the planet today (Losos 2017)?

If we consider Gould's thought experiment and map of diversity through a series of phylogenetic trees (see Fig. 1.7), it is not unreasonable to imagine a period of enormous experimentation during the Cambrian explosion some 541 MYA. At that time there would have been more *space* for experimental solutions, and there is no telling whether vertebrates would have come out successfully, especially when one considers the impressive adaptability and range of solutions among the invertebrates of that period. In fact, the Cambrian explosion marked the early diversification of new lineages associated with substantial experimentation with varied morphologies. Over the previous hundreds of millions of years, this variation has generally been found to decrease through evolutionary time. In other words, the random events of that *explosive* period may have very well set in stone that morphological body plans (i.e., *bauplans*) would dominate the animal kingdom today. Of course, this same line of thought can be applied to other kingdoms and periods of evolutionary time as well.

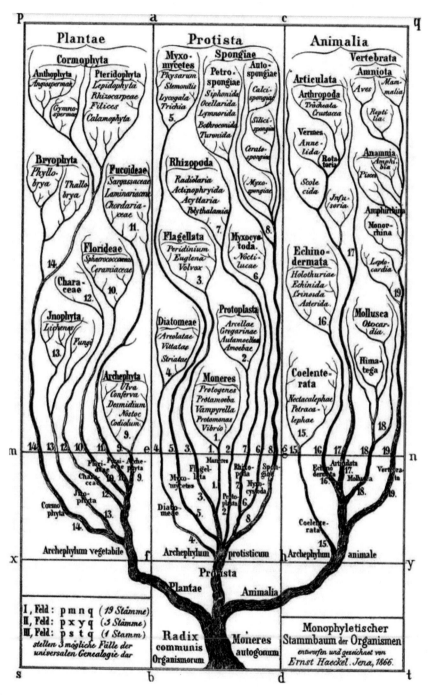

Fig. 1.5 Haeckel's phylogenetic tree, a glimpse into our understanding of the living world in the nineteenth century (Haeckel 1866)

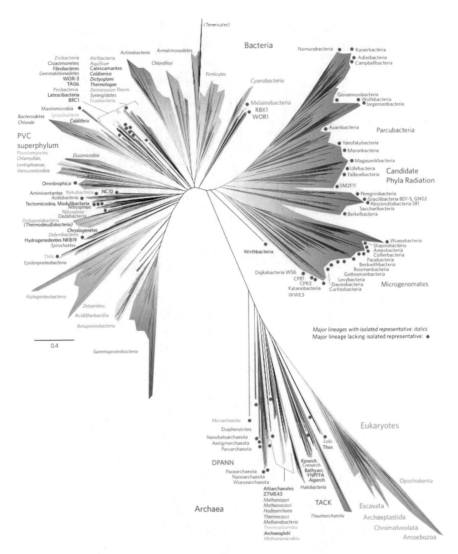

Fig. 1.6 A modern depiction of diversity presented by Hug et al. (2016) in *a new view of the tree of life* that illustrates how small our (human) contribution as Eukaryotes is to the whole of the living world

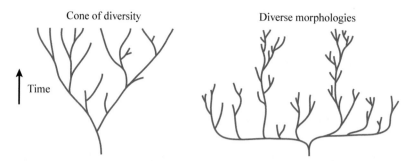

Fig. 1.7 Gould's cone of increasing diversity (left) and a more recent revision of the phylogenetic tree depicting decimation and diversification across numerous morphologies (right). Figure inspired by Gould (1990)

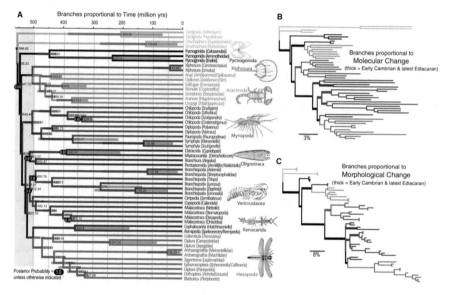

Fig. 1.8 The lineage of a group of invertebrates (reproduced from Lee et al. (2013)) is represented using three different metrics in which the branch thickness is proportional to: (**a**) time; (**b**) molecular change; and (**c**) morphological change. Note the relatively large amount of molecular and morphological change that occurred during the very early stages

On that note, recent advances in genomics have allowed us to look at evolution through a different metric. In Fig. 1.8, we see the phylogenetic tree of a subset of invertebrate evolution but scaled in three different manners (Lee et al. 2013): (A) in time; (B) through molecular change; and (C) through morphological change. When focusing on the thick black lines from the Cambrian explosion, one notes that morphological variation (i.e., evolution) moved forward at a blistering pace in those early years only to slow right down ever since. Although we have always assumed evolution to act at a constant pace, it turns out that rates of evolution can be orders of magnitude faster than predicted through Darwinian evolution and that neo-Lamarckian evolution (hereditary epigenetics), which is governed by heritable changes in gene expression that do not involve changes to DNA, can be used to explain this fascinating process.

1.3.3 Evolutionary Convergence

In contrast to the above arguments on contingency, there is much evidence to support that convergence is working diligently at bringing seemingly disparate systems together in form and function. The classic example that one may use in bio-propulsion would be the convergence of lunate tail–fin shapes seen among four distinct groups: the ichthyosaurs (200 MYA), the lamnid sharks (50–70 MYA), the scombroids (50–70 MYA), and of course cetaceans (40–50 MYA). In some ways, muscle and tissue-based dolphin flukes could not be any more different than bony ichthyosaur tails, in both their evolution and material properties, yet, from a hydrodynamic point of view, these two animals are almost indistinguishable.

Other prominent examples include strong analogies between mammalian placentals and Australian marsupials, for which there are uncanny resemblances in functional morphology. Then, there is the mammalian (human) versus invertebrate (octopus) eye comparison. Their common ancestor goes back as far as the Cambrian explosion, and yet on the surface, these advanced ocular sensors operate in a very similar manner. In fact, despite our typical vertebrate/mammalian/human arrogance, vertebrate eyes have nerves that attach to the front end of the retina's photoreceptors such that nerves bundle together and form a blind spot. This limitation (weakness) is not at all present within our distant octopus relatives as their nerve bundles attach to the back of the retina's photoreceptors, as one would hope for!

Needless to say, finding examples of convergent evolution is bountiful and provides excellent motivation to reverse-engineer their function. However, one must always reflect on constraints before blindly copying. When one finds an example of convergence, there seems to be a counterpoint right around the next corner. For instance, if there are, or at least have been, marsupial analogies to all placentals, then what are we to make of the platypus?

1.3.4 Biomimetics vs. Bioinspiration

"To invent, you need a good imagination and a pile of junk." – Edison

This quote famously captures the essence of *creative* engineering design, where most attempts inevitably fail but allow for iteration by recycling some of the *junk*. Of course, an iterative approach is nothing new but forms, to some extent, the implicit steps taken through Darwinian evolution. However, through the constraints set upon us by *phylogenetic inertia*, the process through which future evolutionary pathways are limited by previous adaptations, we cannot simply *restart* from a clean slate. Just consider the tortuous path taken by the vertebrates from fish, to land-based tetrapods and then back to an aquatic life through the cetaceans. Although no one will dispute the impressive efficiency and maneuverability of a dolphin, is it fair to assume that they represent what we would call a *global optimum* in engineering? For one, swimming performance is just one evolutionary pressure, and besides, perhaps a propeller *is* more efficient, but dolphins were never given the chance to tinker with this proverbial *junk*.

Like in engineering, nature also takes on a *combinatorial* approach. One classic example is that of the jet engine: Three key components include the compressor, the combustion chamber, and the turbine. This engine, enabling transonic and supersonic flight, was invented in parallel via a combinatorial approach by a number of competing teams during the Second World War. By taking technologies developed much earlier during the Industrial Revolution—i.e., compressors that originate from bellows in forges, combustion chambers that were used in locomotive and internal-combustion engines, and finally turbines used for the energy conversion processes—these teams were able to combine and develop rather than start from scratch.

"...an organ originally constructed for one purpose...may be converted into one for a widely different purpose." – Darwin

Gould coined an interesting term to reflect this process in nature, namely *exaptation*, which is essentially the repurposing of one morphological feature (or a combination of features) for a completely different purpose. Perhaps therein lies some of the greatest leaps found in evolution. For instance, how often do we consider that the origins of avian flight, regardless if we believe in their origins through gliding from trees or wing-assisted inclined running (Dececchi et al. 2016), are completely dependent on the advent of feather development? Feathers, made up of a complicated assemblage of fibrous proteins called keratins (the same proteins that made up scales of reptiles), were first and foremost involved in insulation, waterproofing, and reproduction and only later were *co-opted* for flight. What an incredible repurposing when we consider the broad range of flight observed by hummingbirds right up to the albatross!

So where does the above discussion leave us when wearing our engineer's hat? Do we *mimic* (biomimetics) or do we take *inspiration* (bioinspiration) from nature? And when considering Nature's solutions, do we only consider those standing in front of us today, in this fleeting moment of geological time, or do we reconstruct from the past, where more likely than not, numerous solutions were more elegant but just not given the chance to proceed through changing environmental pressures? For instance, is there really a compelling argument to justify the albatross rather than pterosaurs as inspiration when developing high-efficiency aircraft? Or is the reciprocating motion of a dolphin's tail truly more efficient than that of our best propellers today? Beyond the two striking exceptions found in Nature, e.g., the whip-like rotation of eukaryotic *flagellum* and the autorotation of samaras (maple seeds), biological systems have converged on unsteady (reciprocating/undulating) motion.

Whether such seemingly energy-intensive motions, e.g., a periodic acceleration and deceleration, are truly optimal for propulsion, or simply the result of phylogenetic inertia, is a difficult question to answer. In any event, we probably should not constrain our creative energy to direct mimicry of Nature's current solutions, which, for all intents and purposes, are simply *good enough* for the current environmental pressures and as always constrained by the past (phylogenic inertia). Rather, let us consider hybrids by combining solutions found in Nature and engineering—i.e., the German term *Bionik* combines *Biologie* and *Technik*—and use the staggering variation and time-history of Nature's workshop (paleontology) available to us as inspiration.

1.4   Diversity vs. Generality

It is important to remind ourselves that only an estimated 80% of flowering plants are known. And although approximately 95% of bird species are known, only a small fraction of insects and invertebrate animals have yet even been discovered. But let us not stop there. Less than 10% of fungi and less than 1% of microorganisms have been discovered. And to add insult to injury, of all these *known* species, less 10% have even been studied in any depth at all! So what does this mean? Well, we have only touched the tip of the iceberg. The shear act of categorizing diversity like a stamp collection is overwhelming, but to go in detail and understand mechanistic properties of all these species, is at this stage, completely out of consideration.

So what do we do? Wilson (1992) presents a three-pronged approach including: (1) The in-depth examination of *model species*; (2) The discovery/study of the full diversity of life; and (3) The reconstruction of the evolutionary history of each species, known as the *Tree of Life initiative*. Thankfully we are making enormous progress and sequencing the full genomic structure of species at a blistering pace. But does that approach tell us anything about *how* these species may live? Does it shed light on all of their mechanistic richness? By identifying model species in an *ad hoc* manner, are we discovering new and unique solutions? In the end, we must accept that we are very early on in this process. Perhaps we should even rejoice that there is still so much out there yet to be discovered.

1.4.1   Speciation

All it takes is a stroll in the woods, or a day at the beach, and one can easily be overwhelmed with the sheer diversity found across all imaginable scales. One is certainly not exaggerating when describing variation in Nature as truly *incredible*. Given even the slightest opportunity, evolution will assist in systematically filling out (or replacing) complete ecosystems. Whether at the greatest of depths in the oceans, or in the most uninhabitable temperatures on the surface of the Earth, Nature finds a way to adapt. For this reason, is it really so unreasonable to assume that life could not develop and expand in other solar systems?

In contrast to aforementioned concepts and examples on evolutionary convergence, *speciation* represents the process through which populations rapidly evolve to form distinct species better adapted to the current environmental pressures. This concept stems back to Darwin's observations of beak shape from finches on the Galapogos islands (Darwin 1859). Through the lens of engineering design, this process represents a diversification, albeit always constrained by phylogenetic inertia.

Although *phenotypic* variation is genetically assumed as inherited from parent to offspring, as per Darwin's finches, *plasticity* represents the phenomenon where genetically identical organisms produce different phenotypes depending on their current environmental pressures. The effect of plasticity can be observed over very short spans of time (relatively speaking), from one generation to the next, when a species is suddenly exposed to changing environmental pressures. A great example lies within the transition from aquatic to terrestrial life, where the first fish (e.g., *Tiktaalik*) had to adapt from a comfortable life of buoyancy to supporting their weight using pectoral fins (Ahlberg & Clack 2006). We can even witness these changes in a lab environment with extremely adaptable fish, known as *Polypterus*, that can reconfigure rapidly from one generation to the next for life without or with gravity, that is to say a life in water (buoyancy) or on land, respectively.

Indeed, these examples are highly inspiring, both in that we can witness the pace of evolution through rapid adaptation and that the *best* solution is not necessarily the most energetic but rather the one that can pivot to change most effectively. In many engineering applications, the ability to respond robustly to a broad range of conditions is also often more valuable, and hence, we can again take a lesson or two from this nuanced evolutionary process unfolding around us every day.

Finally, in the study of speciation, we often use jargon such as *primitive* or *derived* to describe traits along a branch (or twig) of a phylogenetic tree. This choice of language speaks to a dated, human-centric view of evolution, where we would erroneously assume that extant species are generally more proficient than those that are now probably extinct. This is, of course, fundamentally wrong. Proficiency can only be measured based on the immediate pressures (constraints). As such, we must be careful when misinterpreting these terms. Any biologist will appreciate that this language is an artifact of a bygone era, but we must take care in educating the rest of us when approaching this topic from engineering, medicine, and other related fields. On that note, the act of replaying history and exploring the mechanistic pathways for speciation is in its infancy, but with the advancements of computing power, artificial intelligence, and better modeling, we might even reach a point someday where we can replay time and possibly predict the process of speciation, as science-fiction-like as that might sound.

1.4.2 Allometry

As a counterpoint to the effort in developing non-dimensional groups in fluid dynamics, *allometry* is used as a powerful means to use scaling laws in order to make sense of Nature's endless heap of diversity in function and form. In its simplest form, allometry, also known as allometric scaling, represents the study of relationships between body size, shape, function, etc., and is often expressed in the form of a power law

$$y = ax^b, \tag{1.1}$$

or in log form

$$\log(y) = a\log(x) + \log(b), \tag{1.2}$$

where a and b are the scaling constants that depend on the two measured quantities (or functions). Just look at the convenient log scaling for avian flight from Sullivan et al. (2019) shown in Fig. 1.9, where weight (on the y-axis) is plotted versus wing loading (lift per wing surface area), clearly showing the square-cube law that so often constrains life (and terrestrial gigantism) on the Earth.

Rather than comparing across solutions from seemingly disparate groups, *ontogenetic* allometry provides a perspective on how a species' features change during development from juvenile to adult phases. This process can be particularly insightful when coupled with fluid dynamics since changes in physical scale often come with sudden changes in flow regimes. For instance, let us consider Fig. 1.10,

Fig. 1.9 The relationship between weight and wing loading for a wide range of bird species (green squares) follows a logarithmic trend, taken from Sullivan et al. (2019)

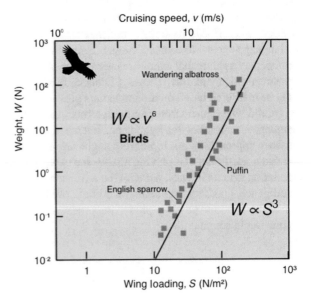

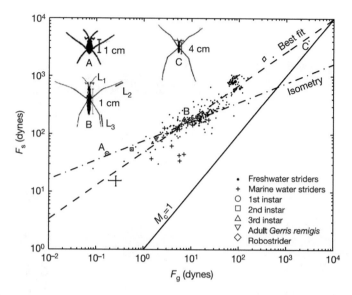

Fig. 1.10 Allometric relation, presented by Hu et al. (2003), between curvature force (F_s) and body weight (F_g) of different species of water striders, including a comparison among juvenile, adult, and even robotic (diamond marker) specimens

where Hu et al. (2003) considered the allometric scaling of water striders and found that the smaller juveniles propel themselves through surface tension, whereas the adults must rely on inertial (vortical) contributions, as we will see in Chap. 3 on Scaling. And of course, such analyses have key implications when designing robots, as seen from the almost anomalous (diamond) point in the top right of Fig. 1.10.

Of course, there are as many ways to view—and therefore scale—a complex set of data as there are parameters themselves. Finding these key parameters will require patience, and finding exceptions to the rule may be just as exciting as seeing all the data collapse onto a single line. Furthermore, when coupled with other tools such as a mapping on a phylogenetic tree, something that is becoming more and more realistic with the pace of sequencing, we can draw even further insight. Consider, for example, the juxtaposition of allometry in hummingbird aerodynamics and their phylogenetic relationship, as shown in Fig. 1.11 (Skandalis et al. 2017). With such richness in data, we might even begin to see how genetics versus plasticity may account for phenotypic expression.

All in all, our efforts to examine diversity in all of its glory, only to try to distill down and consolidate through allometry, moves us conveniently along the path toward abstraction, for which we can then develop simple yet powerful mechanistic descriptions (i.e., models).

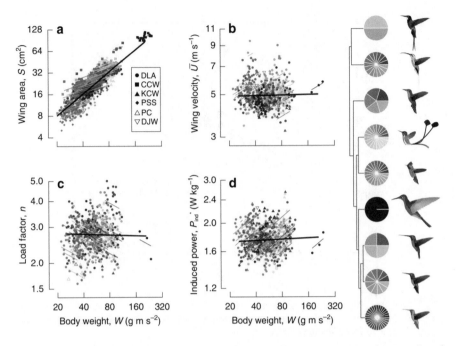

Fig. 1.11 Skandalis et al. (2017) illustrate the allometric divergence among various species of hummingbirds. Note the strong correlation between wing area and body weight for all species, which is absent for each of wing velocity, load factor, and induced power

1.4.3   Abstraction

Absorbing and modeling a seemingly endless amount of complexity, seen through the lens of biological diversity, are certainly no trivial feat. Nevertheless, like in fluid dynamics, where we use scaling arguments equivalent to allometry, there is a method to the madness. In this manner, we can use observations (empiricism) to inform elegant, low-order models, which, in turn, are quite useful in engineering design or perhaps filling in gaps within the sparse fossil record. The process of distilling complexity to a simpler, more convenient form is what we will call *abstraction*, as inspired by the great developments in art over the past centuries.

Of course, there are many forms and levels of abstraction. For instance, a Van Gogh and a Picasso are both forms of abstraction. How approximate should we be in studying a problem at hand? Well, since biological systems are inherently too complex in their pure form, we must seek out compromises. Do we approximate an aorta as a straight pipe? Or do we allow for curvature or perhaps compliance? Do we keep the complex pressure waveform or use a simple sinusoid? Such decisions are tricky and improve with experience. Therefore, developing our ability to use such an approach when tackling new problems will most certainly be a strategy that we will practice throughout this book.

1.5   Observe, Speculate, and Hypothesize

In this book, we focus on developing what may appear as overly simple models of a complex world. However, after *observing* any complex biological process, the next step must always be speculation and the kind of *back-of-the-envelope* calculations that promote discourse. At that point, after a good night's sleep and perhaps a bit more review of theoretical analysis, we are in a position to develop that first key hypothesis. One that can *then* be tested with experiment (be it in a lab with fancy tools or with a big computer). Of course, that first hypothesis will likely be wrong, but having proposed it only to shoot it down, is essential to the productive cycle embedded within the scientific method, as shown in Fig. 1.12.

In the following chapters, we focus on *theory*, accompanied by simple exercises that demonstrate abstraction and some basic working solutions. Sometimes these solutions are trivial, perhaps even glossing over the true beauty of the biological problem in question. However, these exercises allow us to build up our confidence and experience so that later, when tackling a completely *new* problem, we are poised to pass judgment on the pathway forward. Although often it may seem like the development of a simple theoretical model may be a nuisance, or even completely superfluous, there is a danger in skipping this step. With such powerful experimental and computational tools available to us today, it is that much more important that we first consider the underlying physics and abstraction before we bring in the big guns. Once we are neck-deep in data collection—and I mean huge amounts of data!—it might very well be too late.

Finally, this book focuses on establishing theory with occasional application to biomedical, engineering design, and even paleontological problems. We deliberately avoid talk of experimentation and computation, for which these fields continue to advance at a blistering pace. For this reason, the current book does not include citations for all this marvelous work unfolding around us every week. Instead, the reader is encouraged to look at this ever-growing body of literature on biological

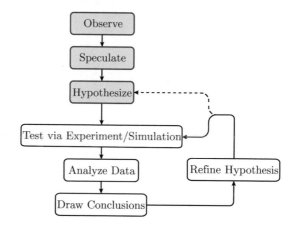

Fig. 1.12 This book focuses on the first three steps of the scientific method, namely observing the natural world, speculating scientific questions based on those observations, and then formulating a hypothesis

topics of their interest, and to use the current book as a springboard toward latter development of complete research (and development) cycles as per the scientific method.

References

Ahlberg, P. E., & Clack, J. A. (2006). A firm step from water to land. *Nature, 440*(7085), 748–749.
Conway Morris, S. (1999). *The crucible of creation: The Burgess Shale and the rise of animals.* Oxford University Press.
Darwin, C. (1859). *On the origins of species.* John Murray.
Dececchi, T. A., Larsson, H. C., & Habib, M. B. (2016). The wings before the bird: An evaluation of flapping-based locomotory hypotheses in bird antecedents. *PeerJ, 4,* e2159.
Gould, S. J. (1990). *Wonderful life: The Burgess shale and the nature of history.* W W Norton & Co.
Haeckel, E. H. P. A. (1866). *Generelle Morphologie der Organismen. Allgemeine Grundzüge der organischen Formen-wissenschaft, mechanisch begründet durch die von Charles Darwin reformierte Descendenztheorie, von Ernst Haeckel.* Reimer, G.
Harvey, W. (1628). Exercitatio anatomica de motu cordis et sanguinis in animalibus. *Annals of Internal Medicine, 74*(6), 1026.
Hu, D. L., Chan, B., & Bush, J. W. M. (2003). The hydrodynamics of water strider locomotion. *Nature, 424*(6949), 663–666.
Hug, L. A., Baker, B. J., Anantharaman, K., Brown, C. T., Probst, A. J., Castelle, C. J., Butterfield, C. N., Hernsdorf, A. W., Amano, Y., Ise, K., Suzuki, Y., Dudek, N., Relman, D. A., Finstad, K. M., Amundson, R., Thomas, B. C., & Banfield, J. F. (2016). A new view of the tree of life. *Nature Microbiology, 1*(5), 16048.
Lee, M. S., Soubrier, J., & Edgecombe, G. D. (2013). Rates of phenotypic and genomic evolution during the Cambrian explosion. *Current Biology, 23*(19), 1889–1895.
Losos, J. B. (2017). *Improbable destinies: Fate, chance, and the future of evolution.* Penguin.
Skandalis, D. A., Segre, P. S., Bahlman, J. W., Groom, D. J. E., Welch, K. C., Witt, C. C., McGuire, J. A., Dudley, R., Lentink, D., & Altshuler, D. L. (2017). The biomechanical origin of extreme wing allometry in hummingbirds. *Nature Communications, 8*(1), 1047.
Sullivan, T. N., Meyers, M. A., & Arzt, E. (2019). Scaling of bird wings and feathers for efficient flight. *Science Advances, 5*(1).
Swammerdam, J. (1950). The "Biblia Naturæ" of Swammerdam. *Nature, 165*(4196), 511–511.
Vesalius, A. (1555). *De humani corporis fabrica* (2nd ed.). Basel.
Wilson, E. O. (1992). *The diversity of life.* W. W. Norton Company.

Chapter 2
Fundamentals

Before we delve into elegant mathematical descriptions and examples of biological flows, and their bio-inspired counterparts, we must first establish the mathematical fundamentals and underlying physical equations needed to proceed. The following sections may represent a form of review for some, whereas a first exposure for others. Regardless of the case, the following tools will be key to reducing the Navier–Stokes equation appropriately so as to draw insight into a given problem of interest.

To start, we must imagine a continuous system in space and time that can be used to characterize our fluid-flow problem. In principle, each parcel of fluid (e.g., molecule) will have its own velocity $\mathbf{v}$ and will move independently as part of the overall fluid field. Effectively, we can imagine an infinitely large number of fluid molecules that need to be accounted for in a given flow. Since it is generally impractical to track the motion of individual molecules, we typically assume a *continuum* such that we can apply field variables defined in time and space, e.g., (x, y, z, t). When the fluid motion is modeled in this way, the information on molecular transport will be lost. However, this macro-modeling of fluid motion reduces the number of equations (needed to describe the motion of each molecule) to a few partial differential equations. In other words, instead of defining the problem using a potentially infinite number of ordinary differential equations (ODEs) describing the motion of molecules, we can reduce the description down to just a few partial differential equations (PDEs). Modeling the equations of motion in fluid dynamics thus requires knowledge of vector calculus, which will be reviewed briefly in the following section.

© The Author(s), under exclusive license to Springer Nature Switzerland AG 2022
D. E. Rival, *Biological and Bio-Inspired Fluid Dynamics*,
https://doi.org/10.1007/978-3-030-90271-1_2

2.1   Vector Calculus

The Nabla ∇ vector operator is based on a combination of partial differential operators. In the Cartesian coordinate system, ∇ is defined as

$$\nabla \equiv \frac{\partial}{\partial x}\hat{i} + \frac{\partial}{\partial y}\hat{j} + \frac{\partial}{\partial z}\hat{k}. \tag{2.1}$$

When ∇ is applied to a scalar field $f(x, y, z)$, it produces a vector that points in the direction of the maximum rate of change of scalar f where:

$$\nabla f = \frac{\partial f}{\partial x}\hat{i} + \frac{\partial f}{\partial y}\hat{j} + \frac{\partial f}{\partial z}\hat{k}. \tag{2.2}$$

This results in a *vector*, where ∇f is referred to as the *gradient* of a scalar function. If ∇ is applied to a vector $\mathbf{F}$ on the other hand, then the result is a tensor referred to as a *gradient tensor*:

$$\nabla \mathbf{F} = \begin{bmatrix} \frac{\partial F_x}{\partial x} & \frac{\partial F_x}{\partial y} & \frac{\partial F_x}{\partial z} \\ \frac{\partial F_y}{\partial x} & \frac{\partial F_y}{\partial y} & \frac{\partial F_y}{\partial z} \\ \frac{\partial F_z}{\partial x} & \frac{\partial F_z}{\partial y} & \frac{\partial F_z}{\partial z} \end{bmatrix}. \tag{2.3}$$

The dot product between ∇ and a vector field $\mathbf{F}$ results in a scalar, where $\nabla \cdot \mathbf{F}$ is known as the *divergence* of a vector field, as shown in Eq. (2.4). This in turn provides a measure of the net flux of $\mathbf{F}$ *out* of a region defined by x, y, z. For example, consider a topological feature known as a source or sink. The divergence of the field determines how much flux points into or out of this region in the flowfield:

$$\nabla \cdot \mathbf{F} = \frac{\partial F_x}{\partial x} + \frac{\partial F_y}{\partial y} + \frac{\partial F_z}{\partial z}. \tag{2.4}$$

The cross product between ∇ and a vector field $\mathbf{F}$ is shown in Eq. (2.5). This results in a *vector*, where $\nabla \times \mathbf{F}$ is referred to as the *curl* of a vector field and determines the local rotation along a specific axis:

$$\begin{aligned} \nabla \times \mathbf{F} &= \left(\frac{\partial}{\partial x}\hat{i} + \frac{\partial}{\partial y}\hat{j} + \frac{\partial}{\partial z}\hat{k} \right) \times \left(F_x\hat{i} + F_y\hat{j} + F_z\hat{k} \right) \\ &= \left(\frac{\partial F_z}{\partial y} - \frac{\partial F_y}{\partial z} \right)\hat{i} + \left(\frac{\partial F_x}{\partial z} - \frac{\partial F_z}{\partial x} \right)\hat{j} + \left(\frac{\partial F_y}{\partial x} - \frac{\partial F_x}{\partial y} \right)\hat{k}. \end{aligned} \tag{2.5}$$

The Laplacian operator ∇^2 (also shown by Δ) is a further extension of the above Del operator, which is defined by a dot product of ∇ on itself. Equation (2.6) shows the application of the Laplacian operator on a scalar field f, which produces a *scalar*

Table 2.1 Tensor ranks of common fluid-dynamic properties

Rank	Form	Fluid property
0	Scalar	p, ρ, T
1	Vector	$\mathbf{v}$
2	N×N matrix	τ_{ij}, ϵ_{ij}

field as well:

$$\nabla \cdot \nabla f = \nabla^2 f = \frac{\partial^2 f}{\partial x^2} + \frac{\partial^2 f}{\partial y^2} + \frac{\partial^2 f}{\partial z^2}. \tag{2.6}$$

If ∇^2 is applied to a vector field the result will be a *vector*:

$$\nabla \cdot \nabla \mathbf{F} = \nabla^2 \mathbf{F} = \nabla^2 F_x \hat{i} + \nabla^2 F_y \hat{j} + \nabla^2 F_z \hat{k}. \tag{2.7}$$

The above operators (∇ and ∇^2) are applied to scalars and vectors, and in many cases the operation results in changing the *rank* of the resulting tensor. The "rank" of a tensor describes its dimension. For example, a scalar and vector are considered zero and first-rank tensors, respectively. The rank of some common fluid properties are given in Table 2.1.

Now, consider an arbitrary fluid element in the flow, as shown in Fig. 2.1, with normal and shear stresses represented by the 3×3 matrix τ_{ij}:

$$\tau_{ij} = \begin{bmatrix} \tau_{xx} & \tau_{xy} & \tau_{xz} \\ \tau_{yx} & \tau_{yy} & \tau_{yz} \\ \tau_{zx} & \tau_{zy} & \tau_{zz} \end{bmatrix}. \tag{2.8}$$

The above stress acting on a fluid element provides a simple example of a rank 2 tensor. We will return to this representation of stress in Sect. 2.3 below.

Tips

The influence of the above vector operators on tensor rank is succinctly reviewed in the following:

- The gradient operator (∇f) changes a scalar (rank 0) into a vector (rank 1);
- The divergence operator reduces rank by 1. For example, the divergence of a tensor ($\nabla \cdot \tau_{ij}$) is a vector, and the divergence of a vector ($\nabla \cdot \mathbf{v}$) is a scalar; and
- The curl of a vector field ($\nabla \times \mathbf{F}$) remains a vector but acts normal to the maximum rotation of the original function.

In Sect. 2.3 we will introduce the Navier–Stokes equation, which in very general terms is used to describe the momentum balance for a given flowfield. This

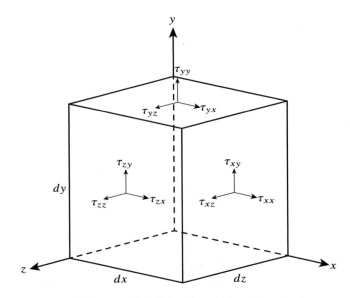

Fig. 2.1 An arbitrary fluid element and its respective normal and shear stresses

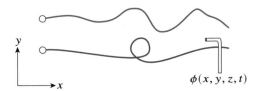

Fig. 2.2 Trajectories of two fluid parcels (red and blue pathlines) and a static probe measuring a local (Eulerian) fluid-flow property

equation contains several spatial and temporal derivatives that are necessary to model fluid motion over time and in space. Here, we will first introduce the two main frameworks (i.e. Eulerian vs. Lagrangian) with which we can describe fluid motions.

Figure 2.2 shows the Lagrangian fluid trajectories (pathlines) and the capture of an Eulerian description of the flow using a static probe. This probe is able to measure some Eulerian property ϕ, such as the velocity vector, pressure, or temperature, at a fixed position. In contrast, in a Lagrangian framework, the fluid property is measured along a fluid pathline over time. Lagrangian descriptions are less commonly used but nevertheless can offer considerable insight into fluid-flow mechanisms, e.g., transport of a given species from one position in space to another. In the current textbook, we will however focus on an Eulerian description.

For a velocity field $\mathbf{v}$, the variation of a fluid-flow property $\phi\,(x, y, z, t)$ at a fixed position can be described by temporal and convective spatial terms, as shown here:

$$d\phi = \underbrace{\frac{\partial \phi}{\partial t} dt}_{\text{temporal}} + \underbrace{\frac{\partial \phi}{\partial x} dx + \frac{\partial \phi}{\partial y} dy + \frac{\partial \phi}{\partial z} dz}_{\text{spatial}}. \tag{2.9}$$

The velocity field $\mathbf{v}$ can be described through infinitesimally small displacements of $dx = u\,dt$, $dy = v\,dt$, and $dz = w\,dt$, where u, v, and w are the velocity components in a Cartesian coordinate system. Substituting these terms for velocity into Eq. (2.9) will result in the following:

$$d\phi = dt\frac{\partial \phi}{\partial t} + u\,dt\frac{\partial \phi}{\partial x} + v\,dt\frac{\partial \phi}{\partial y} + w\,dt\frac{\partial \phi}{\partial z}. \tag{2.10}$$

Now dividing both sides by dt, the expression for temporal variation for any fluid parameter is obtained, often referred to as the *material derivative* and represented by the notation D/Dt:

$$\frac{D\phi}{Dt} = \frac{\partial \phi}{\partial t} + u\frac{\partial \phi}{\partial x} + v\frac{\partial \phi}{\partial y} + w\frac{\partial \phi}{\partial z}, \tag{2.11}$$

where the first term on the right-hand side of Eq. (2.11) describes the variation of the parameter through *time*, whereas the last three terms on the right-hand side indicate how much the parameter of interest changes through *convection*. This can also be represented in vector form:

$$\frac{D\phi}{Dt} = \frac{\partial \phi}{\partial t} + (\mathbf{v} \cdot \nabla)\,\phi. \tag{2.12}$$

Also note that when $\phi = \mathbf{v}$, the material derivative becomes nothing other than the material acceleration:

$$\frac{D\mathbf{v}}{Dt} = \frac{\partial \mathbf{v}}{\partial t} + (\mathbf{v} \cdot \nabla)\,\mathbf{v}. \tag{2.13}$$

Now consider an unusual situation where the measurement probe is capable of following the fluid parcel at the same velocity (i.e., recall Lagrangian framework). This measurement will result in convective terms in Eq. (2.11) to be zero, thus reducing the material derivative to a variation in time only.

Tips

Since the velocity field in Cartesian space can be described through a vector with three components, we obtain $\frac{D\mathbf{v}}{Dt}$ as the acceleration of a local fluid element with three components:

$$\frac{D\mathbf{v}}{Dt} = \frac{Du}{Dt}\hat{i} + \frac{Dv}{Dt}\hat{j} + \frac{Dw}{Dt}\hat{k}.$$

Based on our definition of a material derivative, the acceleration is thus:

$$\frac{D\mathbf{v}}{Dt} = \frac{\partial \mathbf{v}}{\partial t} + (\mathbf{v} \cdot \nabla)\,\mathbf{v}.$$

2.2 Continuity

We will first review the fundamental concept on conservation of mass, also sometimes referred to as *continuity*, using some of the vector-calculus definitions that have been defined earlier. We can then apply what we have learned and explore some interesting features of a typical human heart. Curiously, note that reptilian, avian, and mammalian hearts all share similar left-heart topologies, but we will come to the issue of *scaling* later.

To start, let us remind ourselves of how we pick and define a control volume (CV) so as to most effectively observe or define the passing of a fluid parameter through our problem, as shown in Fig. 2.3.

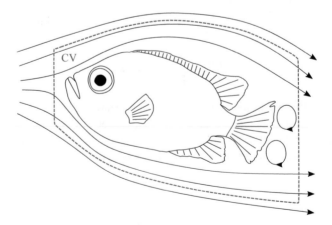

Fig. 2.3 One must pick a control volume (CV) strategically so as to most effectively observe or define a fluid parameter in our system

Fig. 2.4 Defining the parameters of our arbitrary control volume CV

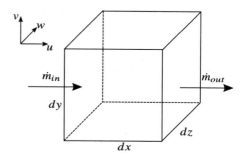

We can then define the accumulation of fluid mass (seawater in this case) within our control volume as follows:

$$\frac{dm_{CV}}{dt} = \sum \dot{m}_{in} - \sum \dot{m}_{out}, \tag{2.14}$$

where the mass flow rate moving in and out of the control volume is defined as

$$\dot{m} = \rho A v(t), \tag{2.15}$$

such that A and v represent the CV surface area and mean velocity, respectively.

We can now rewrite Eq. (2.14) in the following form:

$$\frac{d\rho \Psi}{dt} = \rho A v(t)\Big|_{in} - \rho A v(t)\Big|_{out}, \tag{2.16}$$

and based on the control volume shown in Fig. 2.4 we can define $A_x = dydz$ and $\Psi = dxdydz$, such that we can obtain

$$\frac{d\rho}{dt}dxdydz = \left(\rho u\Big|_x dydz - \rho u\Big|_{x+dx} dydz\right) + \left(\rho v\Big|_y dxdz - \rho v\Big|_{y+dy} dxdz\right)$$

$$+ \left(\rho w\Big|_z dxdy - \rho w\Big|_{z+dz} dxdy\right). \tag{2.17}$$

Note that $\rho u\Big|_x$ represents ρu evaluated on the front face of the CV at position x. In contrast, $\rho u\Big|_{x+dx} = \rho u + \frac{\partial \rho u}{\partial x}dx$ represents the flux on the back face of the CV at position $x + dx$.

We now divide Eq. (2.17) by $\Psi = dxdydz$ throughout, such that we obtain

$$\frac{d\rho}{dt} = \frac{\rho u\Big|_x - \rho u\Big|_{x+dx}}{dx} + \frac{\rho v\Big|_y - \rho v\Big|_{y+dy}}{dy} + \frac{\rho w\Big|_z - \rho w\Big|_{z+dz}}{dz}. \tag{2.18}$$

Now, if we shrink the control volume down to an infinitesimally small fluid element (with differential lengths), i.e., in the limit as $dx, dy, dz \rightarrow 0$, this results in: $\rho u \big|_x - \rho u \big|_{x+dx} \rightarrow -\partial \rho u$, and $dx \rightarrow \partial x$.

Thus, Eq. (2.18) becomes

$$\frac{\partial \rho}{dt} = -\frac{\partial \rho u}{\partial x} - \frac{\partial \rho v}{\partial y} - \frac{\partial \rho w}{\partial z}$$

$$\text{or alternatively} \quad \frac{\partial \rho}{dt} + \frac{\partial \rho u}{\partial x} + \frac{\partial \rho v}{\partial y} + \frac{\partial \rho w}{\partial z} = 0,$$

(2.19)

and in vector form, simply:

$$\frac{\partial \rho}{dt} + \nabla \cdot (\rho \mathbf{v}) = 0. \tag{2.20}$$

When the flow is steady, that is to say the change in mass does not vary in time, continuity becomes

$$\frac{\partial \rho}{dt} = 0 \quad \rightarrow \quad \nabla \cdot (\rho \mathbf{v}) = 0. \tag{2.21}$$

Furthermore, if the flow is incompressible ($\rho = $ const.), the spatial and temporal derivatives of density are zero such that:

$$\frac{\partial \rho}{dt} + \mathbf{v} \cdot \nabla \rho + \rho \nabla \cdot \mathbf{v} = 0 \quad \rightarrow \quad \nabla \cdot \mathbf{v} = 0. \tag{2.22}$$

When we define the material derivative of fluid density (see Eq. (2.12)) as follows:

$$\frac{D\rho}{Dt} = \frac{\partial \rho}{dt} + \mathbf{v} \cdot \nabla \rho, \tag{2.23}$$

we obtain an alternate form to express continuity such that:

$$\frac{D\rho}{Dt} + \rho \nabla \cdot \mathbf{v} = 0. \tag{2.24}$$

In cylindrical coordinates the divergence is

$$\nabla \cdot \mathbf{v} = \frac{1}{r} \frac{\partial}{\partial r} (r v_r) + \frac{1}{r} \frac{\partial v_\theta}{\partial \theta} + \frac{\partial v_z}{\partial z}, \tag{2.25}$$

where r, θ, and z are radial, azimuthal, and axial coordinates such that for a compressible flow can alternatively be expressed as follows:

$$\frac{\partial \rho}{\partial t} + \frac{1}{r}\frac{\partial}{\partial r}(\rho r v_r) + \frac{1}{r}\frac{\partial}{\partial \theta}(\rho v_\theta) + \frac{\partial}{\partial z}(\rho v_z) = 0. \tag{2.26}$$

2.2.1 Exercise with Continuity

Let us now explore conservation of mass through the left side of a beating human heart, i.e., from left atrium (LA), to left ventricle (LV) and then finally out into the aorta, as shown schematically in Fig. 2.5. Again, it is interesting to note here that reptilian, avian, and mammalian hearts share the same common features in the left side of their hearts, whereas amphibians and fish have varying topologies.

As the LV contracts, blood is ejected from the heart into the aorta. The LV then fills and empties again for every contraction. Using conservation of mass, let us calculate $v(t)$, defined as the magnitude of the time-dependent velocity $\mathbf{v}(t)$, at various stages of the heart-beat cycle as shown in Fig. 2.6, at which blood passes through the aortic valve.

During *ejection* (stage B to C in Fig. 2.6), blood flows out of the LV and into the aorta as described by $f(t)$ over 0.05s$\leq t \leq$ 0.30s. For the current example, we have chosen to model $f(t)$ using a power-law relation:

$$f(t) = \Psi_{BC} = \alpha(t - \beta)^{-1/4} \text{ mL}, \tag{2.27}$$

Fig. 2.5 Schematic representation of blood flow in left side of heart

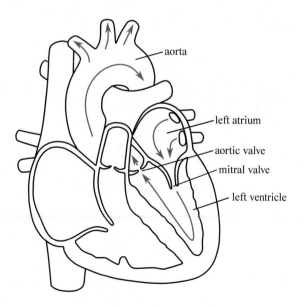

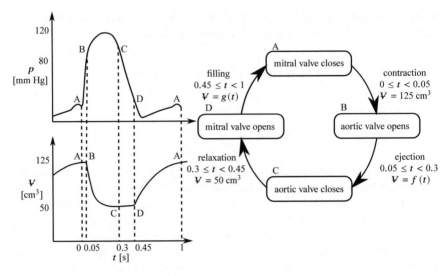

Fig. 2.6 Variation in blood pressure and volume in the left ventricle over various stages of a heart-beat cycle

where α and β are constants that can be calculated with the following input:

$$t = 0.05 \text{ s}, \quad V = 125 \text{ mL}$$
$$t = 0.30 \text{ s}, \quad V = 50 \text{ mL}. \tag{2.28}$$

We now solve for α and β, and obtain values of $\alpha = 35.6$ and $\beta = 0.04$. Thus, we obtain an expression for ejection:

$$V_{BC}(t) = 35.6(t - 0.04)^{-1/4} \text{mL}. \tag{2.29}$$

Now in order to calculate the magnitude of velocity $v(t)$ at the aortic valve, consider that mass must be conserved through the aortic valve. Recall from conservation of mass:

$$\frac{dm}{dt} = \dot{m}_{\text{in}} - \dot{m}_{\text{out}}, \tag{2.30}$$

and with the mitral valve closed ($\dot{m}_{\text{in}} = 0$):

$$\frac{d}{dt}(\rho V) = -\rho A_v v(t). \tag{2.31}$$

If we assume that the aortic-valve's cross section is circular when fully open (i.e. $A_v = \pi R_v^2$) and that blood is incompressible ($\rho = $const.), we obtain

$$\rho \frac{d\forall}{dt} + \forall \frac{d\rho}{dt} = -\rho \pi R_v^2 v(t), \tag{2.32}$$

where R_v is the inner radius of the aorta. For the incompressible condition, this further reduces to

$$\rho \frac{d\forall}{dt} = -\rho \pi R_v^2 v(t), \tag{2.33}$$

$$\frac{d\forall}{dt} = -\pi R_v^2 v(t), \tag{2.34}$$

and substituting in the expression for volume during ejection (Eq. (2.29)):

$$v(t) = -\frac{1}{\pi R_v^2} \frac{d}{dt} \left[35.6(t - 0.04)^{-1/4} \right] \text{ m/s}, \tag{2.35}$$

representing the average velocity over the aortic-valve cross section during ejection (0.05 s$\leq t \leq$ 0.30 s). A similar analysis could be performed for the filling of the left ventricle through the mitral valve by fitting a function for volume $\forall$ over time during stage D to A.

2.3 Navier–Stokes Equation

We can now apply Newton's second law ($m\mathbf{a} = \mathbf{F}$) to any parcel of fluid so long as our assumption of a continuum holds. Let us return to the infinitesimally small volume ($d\forall$) from Fig. 2.1 where:

$$\rho \forall \frac{D\mathbf{v}}{Dt} = \mathbf{F} = \mathbf{F}_{\text{body}} + \mathbf{F}_{\text{surface}}. \tag{2.36}$$

In Eq. (2.36), body forces $\mathbf{F}_{\text{body}}$ arise from external fields such as gravity or applied electromagnetic potential fields. For now it is assumed that body forces are conservative and limited only to gravity, thus resulting in $\mathbf{F}_{\text{body}} = \rho d\forall \mathbf{g}$.

Surface forces ($\mathbf{F}_{\text{surface}}$), on the other hand, are exerted on the fluid element's surfaces; recall Fig. 2.1 and the stress tensor in Eq. (2.8). Note that similar to the strain-rate tensor (ϵ_{ij}) defined below, the stress tensor τ_{ij} is symmetric such that $\tau_{ij} = \tau_{ji}$. Therefore, the total force in each direction, when accounting for all stresses, is

$$dF_x = \tau_{xx}dydz + \tau_{yx}dxdz + \tau_{zx}dxdy$$
$$dF_y = \tau_{xy}dydz + \tau_{yy}dxdz + \tau_{zy}dxdy \qquad (2.37)$$
$$dF_z = \tau_{xz}dydz + \tau_{yz}dxdz + \tau_{zz}dxdy.$$

Let us start by considering a fluid element in equilibrium (i.e. the hydrostatic condition), when all above forces are balanced through equal and opposite forces on the back faces (surfaces) of the fluid element. When the fluid element begins to accelerate, the front- and back-face stresses will differ from one another, e.g., $\tau_{xx} = \tau_{xx,\text{back}} + \frac{\partial \tau_{xx}}{\partial x}dx$. Thus, the net force between the front face and the back face in the x-direction becomes

$$dF_{x,\text{net}} = \left(\frac{\partial \tau_{xx}}{\partial x}dx\right)dydz + \left(\frac{\partial \tau_{yx}}{\partial y}dy\right)dxdz + \left(\frac{\partial \tau_{zx}}{\partial z}dz\right)dxdy. \qquad (2.38)$$

When we examine this force balance on a per unit volume basis, the force in the x-direction becomes

$$f_x = \frac{\partial \tau_{xx}}{\partial x} + \frac{\partial \tau_{xy}}{\partial y} + \frac{\partial \tau_{xz}}{\partial z}, \qquad (2.39)$$

where the indices can flip since $\tau_{ij} = \tau_{ji}$. We note that this result is the same as taking the divergence of the stress tensor, and so for all components:

$$\mathbf{F}_{\text{surface}} = \nabla \cdot \tau_{ij} = \frac{\partial \tau_{ij}}{\partial x_j}, \qquad (2.40)$$

whereby in index notation we define i and j as the free and dummy indices, respectively. Now substituting Eq. (2.40) into Eq. (2.36), Newton's law can be reformulated such that:

$$\rho \frac{D\mathbf{v}}{Dt} = \rho \mathbf{g} + \nabla \cdot \tau_{ij}, \qquad (2.41)$$

where we have only one body force due to gravity.

The following briefly describes how we must express τ_{ij} in terms of the velocity field $\mathbf{v}$ to reduce the number of variables in our equation. In the simplest form, we relate the stress tensor τ_{ij} to the strain-rate tensor ϵ_{ij} through the assumption of a viscous deformation-rate law. Now let us return to our arbitrary fluid element in the flow, as shown in Fig. 2.1, but now describing the individual components of the strain-rate tensor ϵ_{ij} we find

Fig. 2.7 Linear law relating stress with strain rate for a Newtonian fluid (analogous to Hookean elasticity)

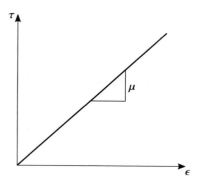

$$\epsilon_{ij} = \begin{bmatrix} \epsilon_{xx} & \epsilon_{xy} & \epsilon_{xz} \\ \epsilon_{yx} & \epsilon_{yy} & \epsilon_{yz} \\ \epsilon_{zx} & \epsilon_{zy} & \epsilon_{zz} \end{bmatrix}. \tag{2.42}$$

The simplest fluids (known as Newtonian fluids) have linear relationships between stress and strain such that, in simple terms, the local stress is proportional to the local strain rate, as shown in Fig. 2.7. This is analogous to the linear relationship known as Hookean elasticity that we use in solid mechanics to describe elastic materials. However, there is an important distinction: In a Newtonian fluid the stress will be proportional to the rate of strain and not just the strain of the fluid element itself.

Stokes (1845) defined three postulates for common Newtonian fluids such as water and air:

- The fluid is continuous and τ_{ij} is a linear function of ϵ_{ij};
- The fluid is isotropic, i.e. fluid properties are independent of direction; and
- When $\epsilon_{ij} = 0$, the force balance reduces to hydrostatic condition $\nabla p = \rho \mathbf{g}$.

With these above criteria, and without concern for its derivation in the context of this textbook, it can be shown that:

$$\tau_{ij} = -p\delta_{ij} + \mu \left(\frac{\partial u_i}{\partial x_j} + \frac{\partial u_j}{\partial x_i} \right) + \delta_{ij} \lambda \nabla \cdot \mathbf{v}, \tag{2.43}$$

where the average compression stress on the fluid element in Cartesian space is

$$p = -\frac{1}{3} \left(\tau_{xx} + \tau_{yy} + \tau_{zz} \right) \tag{2.44}$$

and the Kronecker delta (δ_{ij}) is defined as

$$\delta_{ij} = 1 \text{ if } i = j, \delta_{ij} = 0 \text{ if } i \neq j. \tag{2.45}$$

It also should be noted that the second coefficient of viscosity (λ) can generally be neglected in biological-fluid dynamics problems since this effect is pertinent only in cases with normal shocks and sound-wave absorption attenuation, which are far beyond the scope of the current textbook.

Before substituting the expression for stress in Eq. (2.43) back into Eq. (2.41), we will restrict ourselves to incompressible fluid flows ($\nabla \cdot \mathbf{v} = 0$) with constant viscosity ($\mu \neq f(p, T)$), representative of common Newtonian (linear) fluids. Thus, we obtain the Navier–Stokes equation (for incompressible and Newtonian flows):

$$\rho \frac{D\mathbf{v}}{Dt} = \rho \mathbf{g} - \nabla p + \mu \nabla^2 \mathbf{v} \tag{2.46}$$

where μ, known as the *dynamic viscosity*, represents the coefficient of proportionality between stress and strain rate for a linear fluid. If we expand the material derivative and account for $\rho = const.$, we obtain

$$\frac{\partial \mathbf{v}}{\partial t} + (\mathbf{v} \cdot \nabla)\mathbf{v} = \mathbf{g} - \frac{1}{\rho} \nabla p + v \nabla^2 \mathbf{v}, \tag{2.47}$$

where v is referred to as the *kinematic viscosity* and defined as $v = \frac{\mu}{\rho}$.

Tips

The above form of the Navier–Stokes equation can be solved using p and $\mathbf{v}$ first, and then later (if necessary), we can solve for scalars, such as temperature (T), using the energy equation.

As we have seen in Sect. 2.1, we can expand out to Cartesian space based on the following:

$$\nabla^2 = \frac{\partial^2}{\partial x^2} + \frac{\partial^2}{\partial y^2} + \frac{\partial^2}{\partial z^2}$$

$$\mathbf{v} \cdot \nabla = u \frac{\partial}{\partial x} + v \frac{\partial}{\partial y} + w \frac{\partial}{\partial z} \tag{2.48}$$

and in turn we obtain the following assuming only one body force (gravity) acting in the z-direction:

$$\frac{\partial u}{\partial t} + u\frac{\partial u}{\partial x} + v\frac{\partial u}{\partial y} + \omega\frac{\partial u}{\partial z} = -\frac{1}{\rho}\frac{\partial p}{\partial x} + v\left[\frac{\partial^2 u}{\partial x^2} + \frac{\partial^2 u}{\partial y^2} + \frac{\partial^2 u}{\partial z^2}\right];$$

$$\frac{\partial v}{\partial t} + u\frac{\partial v}{\partial x} + v\frac{\partial v}{\partial y} + \omega\frac{\partial v}{\partial z} = -\frac{1}{\rho}\frac{\partial p}{\partial y} + v\left[\frac{\partial^2 v}{\partial x^2} + \frac{\partial^2 v}{\partial y^2} + \frac{\partial^2 v}{\partial z^2}\right];$$

$$\frac{\partial w}{\partial t} + u\frac{\partial w}{\partial x} + v\frac{\partial w}{\partial y} + \omega\frac{\partial w}{\partial z} = -\frac{1}{\rho}\frac{\partial p}{\partial z} + v\left[\frac{\partial^2 w}{\partial x^2} + \frac{\partial^2 w}{\partial y^2} + \frac{\partial^2 w}{\partial z^2}\right] + g.$$

$$(2.49)$$

In a similar fashion, we can express the Navier–Stokes equation in cylindrical coordinates (r, θ, z)—also with one body force acting in the z-direction—such that for r-momentum:

$$\frac{\partial v_r}{\partial t} + v_r\frac{\partial v_r}{\partial r} + \frac{v_\theta}{r}\frac{\partial v_r}{\partial \theta} + v_z\frac{\partial v_r}{\partial z} - \frac{v_\theta^2}{r}$$

$$= -\frac{\partial p}{\partial r} + v\left[\frac{\partial}{\partial r}\left(\frac{1}{r}\frac{\partial}{\partial r}(rv_r)\right) + \frac{1}{r^2}\frac{\partial^2 v_r}{\partial \theta^2} + \frac{\partial^2 v_r}{\partial z^2} - \frac{2}{r^2}\frac{\partial v_\theta}{\partial \theta}\right],$$

$$(2.50)$$

θ-momentum:

$$\frac{\partial v_\theta}{\partial t} + v_r\frac{\partial v_\theta}{\partial r} + \frac{v_\theta}{r}\frac{\partial v_\theta}{\partial \theta} + v_z\frac{\partial v_\theta}{\partial z} + \frac{v_r v_\theta}{r}$$

$$= -\frac{1}{r}\frac{\partial p}{\partial \theta} + v\left[\frac{\partial}{\partial r}\left(\frac{1}{r}\frac{\partial}{\partial r}(rv_\theta)\right) + \frac{1}{r^2}\frac{\partial^2 v_\theta}{\partial z^2} + \frac{\partial^2 v_\theta}{\partial z^2} + \frac{2}{r^2}\frac{\partial v_r}{\partial \theta}\right],$$

$$(2.51)$$

and z-momentum:

$$\frac{\partial v_z}{\partial t} + v_r\frac{\partial v_z}{\partial r} + \frac{v_\theta}{r}\frac{\partial v_z}{\partial \theta} + v_z\frac{\partial v_z}{\partial z}$$

$$= -\frac{\partial p}{\partial z} + v\left[\frac{1}{r}\frac{\partial}{\partial r}\left(r\frac{\partial v_z}{\partial r}\right) + \frac{1}{r^2}\frac{\partial^2 v_z}{\partial \theta^2} + \frac{\partial^2 v_z}{\partial z^2}\right] + g.$$

$$(2.52)$$

Continuity in cylindrical coordinates takes on the following form:

$$\frac{\partial \rho}{\partial t} + \frac{1}{r}\frac{\partial}{\partial r}(r\rho v_r) + \frac{1}{r}\frac{\partial}{\partial \theta}(\rho v_\theta) + \frac{\partial}{\partial z}(\rho v_z) = 0 \qquad (2.53)$$

and for an incompressible flow (ρ = const.), such that $\nabla \cdot \mathbf{v} = 0$, it reduces further to the following form:

$$\frac{1}{r}\frac{\partial}{\partial r}(rv_r) + \frac{1}{r}\frac{\partial v_\theta}{\partial \theta} + \frac{\partial v_z}{\partial z} = 0. \qquad (2.54)$$

2.4 Strain Rate and Vorticity

As previously shown in Eq. (2.43), the stress tensor can be expressed as a function
of pressure p and velocity $\mathbf{v}$. Based on this simplified relationship, tying shear to
strain rate through a linear relationship (i.e., Newtonian fluid), we can establish the
Navier–Stokes equation for constant density ρ and viscosity v:

$$\frac{\partial \mathbf{v}}{\partial t} + (\mathbf{v} \cdot \nabla) \, \mathbf{v} = \mathbf{g} - \frac{1}{\rho} \nabla p + v \nabla^2 \mathbf{v}. \tag{2.47}$$

Tips
The following terms are present in the Navier–Stokes equation:

- $\frac{\partial \mathbf{v}}{\partial t} \rightarrow$ local acceleration (unsteady term);
- $(\mathbf{v} \cdot \nabla) \, \mathbf{v} \rightarrow$ convection (non-linear term);
- $\mathbf{g} \rightarrow$ gravitational acceleration;
- $\frac{1}{\rho} \nabla p \rightarrow$ pressure gradient; and
- $v \nabla^2 \mathbf{v} \rightarrow$ viscous diffusion (2nd-order term).

For cases where the flow is steady and viscosity plays a secondary role, $\frac{\partial \mathbf{v}}{\partial t} = 0$
and $v \nabla^2 \mathbf{v}$ becomes negligible, respectively. In turn, we obtain the Euler equation
(Euler 1757):

$$(\mathbf{v} \cdot \nabla) \, \mathbf{v} = \mathbf{g} - \frac{1}{\rho} \nabla p. \tag{2.55}$$

As shown in Fig. 2.8, we can integrate the Euler equation along a streamline from
points A to B such that:

$$\left(p + \frac{1}{2} \rho V^2 + \rho g y \right)_A = \left(p + \frac{1}{2} \rho V^2 + \rho g y \right)_B, \tag{2.56}$$

so as to obtain the formulation for the Bernoulli equation, which we will revisit later.

When considering the inviscid region away from the body, only normal stresses
exist such that $\tau_{ij} = -p\delta_{ij}$, as indicated in Eq. (2.43). The stress tensor τ_{ij} contains
normal and shear stresses, but we know that a fluid element's rate of deformation can
be broken into four characteristic motions: translation, rotation, extensional strain
(dilatation), and shear strain, as shown in Fig. 2.9.

For a rotational motion, we define an angular rotation on a fluid element as
follows:

$$d\Omega_z = \frac{1}{2} (d\alpha - d\beta), \tag{2.57}$$

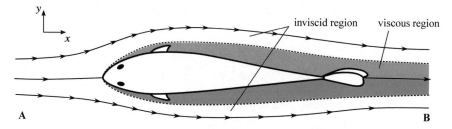

Fig. 2.8 Identification of viscous and inviscid regions around a swimming body such as a fish. Note that within the inviscid region we can integrate along a streamline via Bernoulli. Also consider the *rotational* flow in the viscous regions along the body where red and blue represent clockwise and counterclockwise rotation (vorticity), respectively

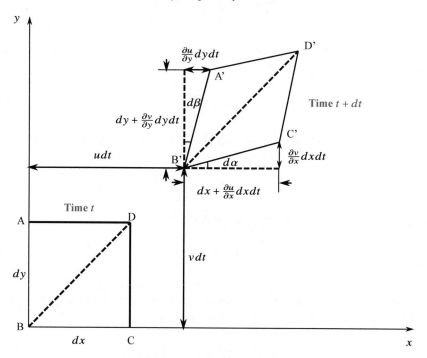

Fig. 2.9 Fluid-element deformation through four characteristic motions: translation, rotation, extensional strain (dilatation), and shear strain

where

$$d\alpha = \lim_{dt \to 0} \left(\tan^{-1} \frac{\frac{\partial v}{\partial x} dx dt}{dx + \frac{\partial u}{\partial x} dx dt} \right) = \frac{\partial v}{\partial x} dt$$

$$d\beta = \lim_{dt \to 0} \left(\tan^{-1} \frac{\frac{\partial u}{\partial y} dy dt}{dy + \frac{\partial v}{\partial y} dy dt} \right) = \frac{\partial u}{\partial y} dt.$$

(2.58)

Then, substituting Eq. (2.58) into Eq. (2.57), we obtain the z-component of the angular rotation rate:

$$\frac{d\Omega_z}{dt} = \frac{1}{2}\left(\frac{\partial v}{\partial x} - \frac{\partial u}{\partial y}\right), \tag{2.59}$$

and similarly, for $x-$ and $y-$components we are able to identify the following angular rotation rates:

$$\frac{d\Omega_x}{dt} = \frac{1}{2}\left(\frac{\partial w}{\partial y} - \frac{\partial v}{\partial z}\right)$$
$$\frac{d\Omega_y}{dt} = \frac{1}{2}\left(\frac{\partial u}{\partial z} - \frac{\partial w}{\partial x}\right). \tag{2.60}$$

In this way, we are able to relate a new quantity called *vorticity* to the angular rotation rate simply as

$$\boldsymbol{\omega} = 2\frac{d\boldsymbol{\Omega}}{dt}. \tag{2.61}$$

Vorticity is defined as the curl of the velocity field, representing local rotation in a fluid flow, and is written as follows:

$$\boldsymbol{\omega} = \nabla \times \mathbf{v}, \tag{2.62}$$

where the vorticity vector is *solenoidal* such that

$$\nabla \cdot \boldsymbol{\omega} = 0. \tag{2.63}$$

For extensional strain, where the fluid-element's volume changes in time, we will save such a discussion for a future course on compressible flows. Finally, the strain rate ϵ_{ij} can be recovered in a similar manner to the rotation rate:

$$\epsilon_{xy} = \frac{1}{2}\left(\frac{d\alpha}{dt} + \frac{d\beta}{dt}\right) = \frac{1}{2}\left(\frac{\partial v}{\partial x} + \frac{\partial u}{\partial y}\right), \tag{2.64}$$

and similarly,

$$\epsilon_{yz} = \frac{1}{2}\left(\frac{\partial w}{\partial y} + \frac{\partial v}{\partial z}\right);$$
$$\epsilon_{zx} = \frac{1}{2}\left(\frac{\partial u}{\partial z} + \frac{\partial w}{\partial x}\right). \tag{2.65}$$

Recall that like shear $\tau_{ij} = \tau_{ji}$, the strain-rate tensor is also symmetric such that $\epsilon_{ij} = \epsilon_{ji}$.

By combining Eq. (2.43) with Eqs. (2.59) and (2.64), it now becomes possible for each component of the velocity gradient to be resolved into a strain rate (symmetric) plus an angular velocity (antisymmetric):

$$\frac{\partial u_i}{\partial x_j} = \epsilon_{ij} + \frac{d\Omega_{ij}}{dt}, \tag{2.66}$$

where ϵ_{ij} is the strain rate responsible for viscous stress (losses), and in contrast, $\frac{d\Omega_{ij}}{dt}$ represents the rotation of the fluid element with zero distortion.

Since *vortices* (regions of concentrated vorticity) are generally responsible for mass, momentum, and energy transport, it is often useful to consider vorticity transport *in lieu* of the Navier–Stokes equation. By starting with Eq. (2.47), we replace the convective term using the following vector-calculus identity:

$$(\mathbf{v} \cdot \nabla) \mathbf{v} = \nabla \left(\frac{1}{2} \mathbf{v} \cdot \mathbf{v} \right) - \mathbf{v} \times (\nabla \times \mathbf{v}), \tag{2.67}$$

such that Eq. (2.47) becomes

$$\frac{\partial \mathbf{v}}{\partial t} + \nabla \left(\frac{1}{2} \mathbf{v} \cdot \mathbf{v} \right) - \mathbf{v} \times (\nabla \times \mathbf{v}) = \mathbf{g} - \frac{1}{p} \nabla p + v \nabla^2 \mathbf{v}. \tag{2.68}$$

We then take the curl of the Navier–Stokes equation:

$$\frac{\partial \boldsymbol{\omega}}{\partial t} - \nabla \times (\mathbf{v} \times \boldsymbol{\omega}) = v \nabla^2 \boldsymbol{\omega}. \tag{2.69}$$

Tips
The following helps us simplify:

- The curl of the gradient of any scalar is zero; and
- The curl of a conservative body force such as **g** will also vanish.

We then use another vector identity to expand the second term in Eq. (2.69):

$$\nabla \times (\mathbf{v} \times \boldsymbol{\omega}) = \mathbf{v} (\nabla \cdot \boldsymbol{\omega}) - \boldsymbol{\omega} (\nabla \cdot \mathbf{v}) - (\mathbf{v} \cdot \nabla) \boldsymbol{\omega} + (\boldsymbol{\omega} \cdot \nabla) \mathbf{v}$$
$$= - (\mathbf{v} \cdot \nabla) \boldsymbol{\omega} + (\boldsymbol{\omega} \cdot \nabla) \mathbf{v}. \tag{2.70}$$

To simplify, consider when a vector field, such as $\boldsymbol{\omega}$, only has a vector potential component (the solenoidal field is $\boldsymbol{\omega} = \nabla \times \mathbf{v}$), then the divergence of the vector is zero ($\nabla \cdot \boldsymbol{\omega} = 0$). Furthermore, for an incompressible flow, the divergence of the velocity field also is zero ($\nabla \cdot \mathbf{v} = 0$). Thus, when substituting Eq. (2.70) into

Eq. (2.69), we obtain the *vorticity-transport* equation, also sometimes referred to as the Helmholtz equation of hydrodynamics:

$$\frac{\partial \boldsymbol{\omega}}{\partial t} + (\mathbf{v} \cdot \nabla)\,\boldsymbol{\omega} = (\boldsymbol{\omega} \cdot \nabla)\,\mathbf{v} + v\nabla^2\boldsymbol{\omega}. \tag{2.71}$$

Tips

Pressure no longer appears explicitly in the vorticity-transport equation. Instead we are left with the following terms:

- $\frac{\partial \boldsymbol{\omega}}{\partial t} \to$ unsteady term;
- $(\mathbf{v} \cdot \nabla)\,\boldsymbol{\omega} \to$ convection term;
- $(\boldsymbol{\omega} \cdot \nabla)\,\mathbf{v} \to$ stretching and tilting term (only in three-dimensional case); and
- $v\nabla^2\boldsymbol{\omega} \to$ diffusion term.

2.4.1 Exercise with Vorticity

Let us consider the periodic expulsion of blood from the left ventricle through the aortic valve. We wish to estimate the vorticity field within the ascending aorta just after the valve opens. We determine an expression for the velocity, $\mathbf{v}(t) = \frac{89}{10\pi R_v^2}(t - 0.04)^{-5/4}$, describing blood ejected through the aortic valve. At the instant the valve opens ($t \approx 0.05$s), a jet is formed at the entrance to the aorta. Let us assume that the jet has a top-hat function, as shown in Fig. 2.10, with a peak velocity equal to $\mathbf{v}(t)$ evaluated at $t \approx 0.05$s.

We can first calculate the curl of the velocity field (the vorticity field) in the aorta when the aortic valve opens (assume the vessel is cylindrical). Afterwards we can then explore what this quantity may represent physically.

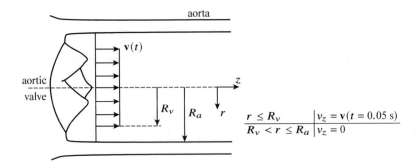

Fig. 2.10 Side view of aortic valve open with top-hat velocity profile

We assume $\mathbf{v} = v_z \hat{k}$ and $v_r = v_\theta = 0$ using cylindrical coordinates such that:

$$v_z(t = 0.05\text{s}) = \frac{89}{10\pi R_v^2}(0.05 - 0.04)^{-5/4}\text{m/s} = \text{const.} \qquad (2.72)$$

Several terms with the curl of the velocity field (vorticity field) vanish:

$$\nabla \times \mathbf{v} = \frac{1}{r}\left[\frac{\partial v_z}{\partial \theta} - r\frac{\cancel{\partial v_\theta}}{\cancel{\partial z}}\right]\hat{r} + \left[\frac{\cancel{\partial v_r}}{\cancel{\partial z}} - \frac{\partial v_z}{\partial r}\right]\hat{\theta} + \frac{1}{r}\left[r\frac{\cancel{\partial v_\theta}}{\cancel{\partial r}} - \frac{\cancel{\partial v_r}}{\cancel{\partial \theta}}\right]\hat{k}. \qquad (2.73)$$

We also observe that on the flat velocity profiles:

$$\nabla \times \mathbf{v} = \begin{cases} 0 & r \leq R_v \\ 0 & R_v < r \leq R_a \end{cases} \qquad (2.74)$$

such that at $r = R_v$, the velocity $\mathbf{v}$ is discontinuous and steps from 0 to v_z, as shown in Fig. 2.11. Across this discontinuity the curl of the velocity tends to infinity (we know this is only a crude approximation at this stage). Nevertheless, the curl of the velocity field (vorticity $\boldsymbol{\omega}$) helps us identify regions of strong rotational motion in the flow.

In Fig. 2.12 we observe zero vorticity throughout except for singularities at $r = \pm R_v$.

When the left ventricle has finished contracting the aortic valve closes and the jetting of blood entering the aorta ends. The closure causes a ring of concentrated vorticity to roll up and advect downstream, as shown in Fig. 2.13.

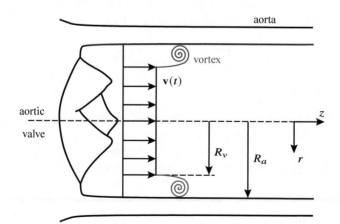

Fig. 2.11 Sharp discontinuity in the top-hat velocity profile is a first crude approximation to reality in which we ignore viscous diffusion

Fig. 2.12 Singularity of
vorticity at $r = \pm R_v$

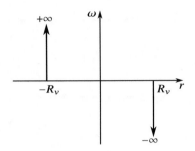

Fig. 2.13 Isometric view of
vortex ring traveling along
aorta

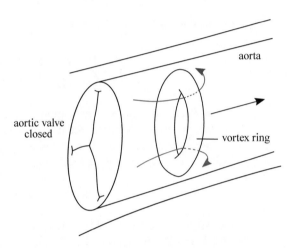

Finally, by interpreting the terms of the incompressible vorticity-transport equation, $D\boldsymbol{\omega}/Dt$, let us discuss the outcome of the vortex ring as it moves along the aorta. The vorticity-transport equation in compact form is as follows:

$$\frac{D\boldsymbol{\omega}}{Dt} = (\boldsymbol{\omega} \cdot \nabla)\mathbf{v} + v\nabla^2\boldsymbol{\omega}. \tag{2.75}$$

We observe the following:

1. The material derivative $D\boldsymbol{\omega}/Dt$ describes how vorticity is transported by the bulk flow, i.e., through acceleration and convection along the aorta.
2. $(\boldsymbol{\omega} \cdot \nabla)\mathbf{v}$ represents how the vortex ring is stretched by the velocity-gradient field. Curvature in the aortic vessel, or strong velocity gradients themselves, can cause the vortex ring to tilt relative to the artery's axis. In order to conserve angular momentum the vortex ring stretches along this new tilted plane but does not grow; as shown in Fig. 2.14.
3. $v\nabla^2\boldsymbol{\omega}$ accounts for the diffusion of vorticity, which is responsible for spreading vorticity spatially over time. Instabilities in the vortex ring, through interactions with the aortic wall, will cause vortex-ring breakup.

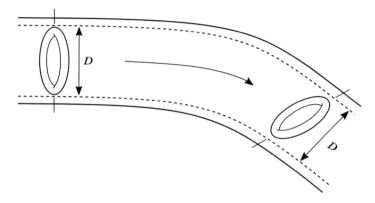

Fig. 2.14 Hypothetical vortex-ring tilting due to aorta curvature

2.5 Circulation and Model Vortices

Now that we have introduced the definition of vorticity,

$$\boldsymbol{\omega} = \nabla \times \mathbf{v}, \tag{2.62}$$

let us see how we can use this feature to describe a series of different vortex structures (vortices) commonly found in biological and bio-inspired flows. For instance, when we identify regions in a fluid flow with concentrated vorticity, we can also describe the *strength* of this rotational region (vortex) in terms of its circulation Γ:

$$\Gamma = \oint \mathbf{v} \cdot d\boldsymbol{\ell}, \tag{2.76}$$

where $d\boldsymbol{\ell}$ is an arbitrarily chosen path enclosing the vortex of interest.

Circulation can be related to the vorticity field using Stokes' Theorem, which is defined as follows:

$$\oint_{\ell} \mathbf{F} \cdot d\boldsymbol{\ell} = \int_{A} (\nabla \times \mathbf{F}) \cdot \mathbf{n} dA \tag{2.77}$$

where for any (3D) surface A, ℓ represents the contour. As such circulation can therefore be defined as follows:

$$\Gamma = \oint \mathbf{v} \cdot d\boldsymbol{\ell} = \int (\nabla \times \mathbf{v}) \cdot \mathbf{n} dA \quad \Rightarrow \quad \Gamma = \int \boldsymbol{\omega} \cdot \mathbf{n} dA. \tag{2.78}$$

Thus, the strength of a given vortex is defined by its circulation, which simply represents the amount of vorticity distributed within its immediate vicinity.

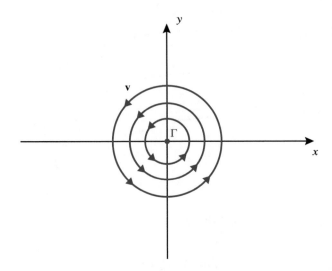

Fig. 2.15 A potential vortex with circulation Γ concentrated at center

Now let us consider some simple models of vortices. Starting with the simplest case of a *potential* vortex as shown in Fig. 2.15. Here, the velocity field can be represented as concentric rings around the point vortex with strength Γ:

$$\mathbf{v} = v_\theta \hat{\theta}, \tag{2.79}$$

and the velocity decays with one over the radius r such that

$$v_\theta = \frac{\Gamma}{2\pi r}. \tag{2.80}$$

We can then check that this particular flow field satisfies the Navier–Stokes equation (θ-momentum). Let us examine the vorticity field in cylindrical coordinates:

$$\boldsymbol{\omega} = \nabla \times \mathbf{v} = \left(\frac{1}{r}\frac{\partial v_z}{\partial \theta} - \frac{\partial v_\theta}{\partial z}\right)\hat{r} + \left(\frac{\partial v_r}{\partial z} - \frac{\partial v_z}{\partial r}\right)\hat{\theta} + \frac{1}{r}\left(\frac{\partial (r v_\theta)}{\partial r} - \frac{\partial v_r}{\partial \theta}\right)\hat{k}$$

$$= \frac{1}{r}\frac{\partial}{\partial r}\left[r\frac{\Gamma}{2\pi r}\right]\hat{k}$$

$$\boldsymbol{\omega} = 0 \quad \rightarrow \quad \text{irrotational.}$$

$$\tag{2.81}$$

Note that this vortex structure is also sometimes referred to as a line or free vortex and has $\omega_z = 0$ everywhere except on a line at the origin where $\omega_z = \infty \hat{k}$, as shown in Fig. 2.16.

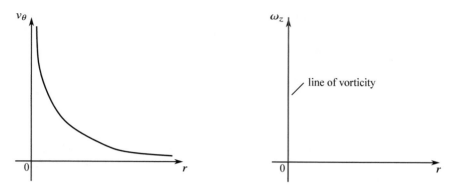

Fig. 2.16   Velocity and vorticity distribution in a potential vortex

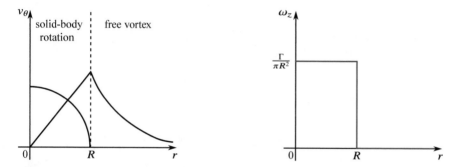

Fig. 2.17   Velocity and vorticity distribution in a Rankine vortex

However, in a more realistic (viscous) fluid medium a discontinuity, such that $v_\theta = \infty$ at its origin, is unrealistic. Viscous diffusion comes into play even at the highest Reynolds numbers, thus generating a core dominated by viscous effects. To account for these viscous effects, a modification to the potential vortex model was proposed. This modification is referred to as a *Rankine* vortex, as shown in Fig. 2.17. In this model we assume that the inner core (of radius R) rotates in solid-body rotation such that:

$$v_\theta(r) = \begin{cases} \frac{\Gamma r}{2\pi R^2}, & r \leq R \quad \text{solid-body rotation (linear)} \\ \frac{\Gamma}{2\pi r}, & r > R \quad \text{free vortex} \end{cases} \tag{2.82}$$

$$\boldsymbol{\omega} = \nabla \times \mathbf{v} = \begin{cases} \frac{1}{r}\frac{\partial}{\partial r}\left[r\frac{\Gamma r}{2\pi R^2}\right]\hat{k} = \frac{\Gamma}{\pi R^2}\hat{k}, & r \leq R \quad (= \text{const.}) \\ 0, & r > R. \end{cases} \tag{2.83}$$

Finally, let us now consider a final, unsteady case, referred to as a Lamb–Oseen vortex (Oseen 1912), in which a plane vortex diffuses radially outward in time. The velocity distribution is given by

$$v_r = 0,$$
$$v_\theta = \frac{\Gamma}{2\pi r} \left[1 - e^{-r^2/4tv} \right]. \tag{2.84}$$

We will start by checking that the above expression is in fact an exact solution to continuity and the Navier–Stokes equation. The following are our key vector operations in cylindrical coordinates:

$$\nabla \cdot \mathbf{v} = \frac{1}{r} \frac{\partial}{\partial r} (r v_r) + \frac{1}{r} \frac{\partial}{\partial \theta} (v_\theta) + \frac{\partial}{\partial z} (v_z)$$

$$\nabla \times \mathbf{v} = \left(\frac{1}{r} \frac{\partial v_z}{\partial \theta} - \frac{\partial v_\theta}{\partial z} \right) \hat{r} + \left(\frac{\partial v_r}{\partial z} - \frac{\partial v_z}{\partial r} \right) \hat{\theta} + \frac{1}{r} \left(\frac{\partial (r v_\theta)}{\partial r} - \frac{\partial v_r}{\partial \theta} \right) \hat{k}, \tag{2.85}$$

along with the θ-component of the Navier–Stokes equation:

$$\frac{\partial v_\theta}{\partial t} + (\mathbf{v} \cdot \nabla) v_\theta + \frac{v_r v_\theta}{r} = -\frac{1}{\rho r} \frac{\partial p}{\partial \theta} + g_\theta + v \left(\nabla^2 v_\theta + \frac{2}{r^2} \frac{\partial v_r}{\partial \theta} - \frac{v_\theta}{r^2} \right), \tag{2.86}$$

where

$$\mathbf{v} \cdot \nabla = v_r \frac{\partial}{\partial r} + \frac{1}{r} v_\theta \frac{\partial}{\partial \theta} + v_z \frac{\partial}{\partial z} \tag{2.87}$$

and

$$\nabla^2 = \frac{1}{r} \frac{\partial}{\partial r} \left(r \frac{\partial}{\partial r} \right) + \frac{1}{r^2} \frac{\partial^2}{\partial \theta^2} + \frac{\partial^2}{\partial z^2}. \tag{2.88}$$

Note that the r- and z-components of momentum each go to zero. First, checking continuity we obtain

$$\nabla \cdot \mathbf{v} = \frac{1}{r} \frac{\partial}{\partial r} (v_r r) + \frac{1}{r} \frac{\partial v_\theta}{\partial \theta} + \frac{\partial v_z}{\partial z} = 0$$
$$\frac{1}{r} \frac{\partial}{\partial \theta} \left[\frac{K}{r} \left[1 - e^{-r^2/4tv} \right] \right] = 0, \tag{2.89}$$

where $K = \Gamma/2\pi$ represents the strength of the vortex. Subsequently, we can also check the Navier–Stokes equation (θ-component) where again most terms are zero

$(v_r = 0, v_z = 0)$ and, with axial symmetry, pressure does not vary in the azimuthal direction either $(\partial p / \partial \theta = 0)$:

$$\frac{\partial v_\theta}{\partial t} + \cancel{v_r \frac{\partial v_\theta}{\partial r}} + \frac{1}{r} v_\theta \cancel{\frac{\partial v_\theta}{\partial \theta}} + \cancel{v_z \frac{\partial v_\theta}{\partial z}} = v \left[\frac{1}{r} \frac{\partial}{\partial r} (r \frac{\partial v_\theta}{\partial r}) + \frac{1}{r^2} \cancel{\frac{\partial^2 v_\theta}{\partial \theta^2}} + \cancel{\frac{\partial^2 v_\theta}{\partial z^2}} - \frac{v_\theta}{r^2} \right]$$

$$\frac{\partial v_\theta}{\partial t} = v \left[\frac{1}{r} \frac{\partial}{\partial r} (r \frac{\partial v_\theta}{\partial r}) - \frac{v_\theta}{r^2} \right]. \tag{2.90}$$

The left-hand side (LHS) of the expression reduces to

$$\frac{\partial}{\partial t} \left[\frac{K}{r} \left[1 - e^{-r^2/4tv} \right] \right] = -\frac{Kr}{4t^2 v} e^{-r^2/4tv}, \tag{2.91}$$

whereas the right-hand side (RHS) can be simplified as follows:

$$\frac{v}{r} \frac{\partial}{\partial r} \left(r \frac{\partial}{\partial r} \left[\frac{K}{r} (1 - e^{-r^2/4tv}) \right] \right) - \frac{v}{r^2} \cdot \frac{K}{r} (1 - e^{-r^2/4tv})$$

$$= \frac{v}{r} \frac{\partial}{\partial r} \left(r \left[\frac{K}{r} \left(\frac{2r}{4tv} e^{-r^2/4tv} \right) - \frac{K}{r^2} \left(1 - e^{-r^2/4tv} \right) \right] \right) - \frac{vK}{r^3} (1 - e^{-r^2/4tv})$$

$$= \frac{v}{r} \frac{\partial}{\partial r} \left[\frac{Kr}{2tv} e^{-r^2/4tv} - \frac{K}{r} (1 - e^{-r^2/4tv}) \right] - \frac{v v_\theta}{r^2}$$

$$= \frac{v}{r} \left[\frac{Kr}{2tv} (-\frac{2r}{4tv}) e^{-r^2/4tv} + \frac{K}{r^2} (1 - e^{-r^2/4tv}) \right] - \frac{v v_\theta}{r^2} = -\frac{Kr}{4t^2 v} e^{-r^2/4tv}. \tag{2.92}$$

We therefore show that the solutions to the LHS and RHS in the Navier–Stokes equation are equal.

Let us now consider how the vorticity

$$\boldsymbol{\omega} = \nabla \times \mathbf{v} = \frac{1}{r} \frac{\partial}{\partial r} (r v_\theta) \hat{k} \tag{2.93}$$

varies over time such that

$$\boldsymbol{\omega} = \frac{1}{r} \frac{\partial}{\partial r} \left[K (1 - e^{-r^2/4tv}) \right] \hat{k}$$

$$= \frac{1}{r} \left[\frac{2rK}{4tv} e^{-r^2/4tv} \right] \hat{k}, \tag{2.94}$$

or simply:

$$\omega_z = \frac{\Gamma}{4\pi t v} e^{-r^2/4tv}. \tag{2.95}$$

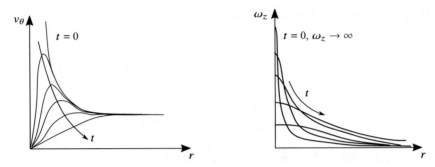

Fig. 2.18 Distribution of velocity and vorticity in a Lamb–Oseen vortex over time

Therefore, at time $t = 0$ we start with a potential vortex (as described above), followed by vorticity diffusing radially outwards over time. This variation in distribution is clearly shown in Fig. 2.18.

Reference

Oseen, C. W. (1912). Über die Wirbelbewegung in einer reibenden Flussigkeit. *Arkiv für Mathematik, Astronomie und Physik, 7*, 14–26.
Stokes, G. G. (1845). On the theories of the internal friction of fluids in motion and of the equilibrium and motion of elastic solids. *Transactions of the Cambridge Philosophical Society, 8*, 287–319.
Euler, L. (1757). Principes généraux de l'etat d'equilibre des fluides. *Mémoires de l'académie des sciences de Berlin, 11*, 217–273.

Chapter 3
Scaling

Now that we have established the foundational tools, the objective of the current chapter will be to explore the breadth of scales, and the stark physical changes that occur, across the vast range of biological and bio-inspired systems. We start by normalizing the Navier–Stokes equation so that we can develop a physical understanding of its various terms. In particular, we see how the Navier–Stokes equation can be manipulated and even simplified for various cases at hand, thus helping us draw insight into key physical processes. By understanding these physical processes we are then able to tackle (and simplify) seemingly complex biological problems ranging from the very tiniest, such as for cellular motility, to the very largest, such as for the drag produced in the turbulent boundary layers developing on our mammalian cousins of the sea.

3.1 Normalization

Before we explore solutions at various biological scales, we must first return to the Navier–Stokes equation. We will show now that for any given problem the flow physics can be cast into a more efficient form through the initial process of *non-dimensionalizing* the Navier–Stokes equation. Since we can rely on some prior experience on dimensional analysis or *similitude* (Buckingham 1915), we take the original form of our function:

$$\rho \frac{\partial \mathbf{v}}{\partial t} + \rho \left(\mathbf{v} \cdot \nabla \right) \mathbf{v} = \rho \mathbf{g} - \nabla p + \mu \nabla^2 \mathbf{v}, \qquad (3.1)$$

and choose appropriate scaling parameters such as:

- L is the characteristic length $\{L\}$;
- U_∞ is the characteristic speed $\{LT^{-1}\}$;

© The Author(s), under exclusive license to Springer Nature Switzerland AG 2022
D. E. Rival, *Biological and Bio-Inspired Fluid Dynamics*,
https://doi.org/10.1007/978-3-030-90271-1_3

- f is the characteristic frequency $\{T^{-1}\}$;
- $p_o - p_\infty$ is the reference pressure difference $\{ML^{-1}T^{-2}\}$; and
- g is the gravitational acceleration $\{LT^{-2}\}$,

where M, L, and T represent dimensions of mass, length, and time, respectively.

We now define non-dimensional variables using the above scaling parameters:

$$t^* = ft \quad, \mathbf{v}^* = \frac{\mathbf{v}}{U_\infty} \quad, \mathbf{g}^* = \frac{\mathbf{g}}{g} \quad, p^* = \frac{p-p_\infty}{p_o-p_\infty} \quad, \text{and } \nabla^* = L\nabla,$$

and substitute t^*, $\mathbf{v}^*$, $\mathbf{g}^*$, p^*, and ∇^* into Eq. (3.1) such that:

$$\rho U_\infty f \frac{\partial \mathbf{v}^*}{\partial t^*} + \frac{\rho U_\infty^2}{L}\left(\mathbf{v}^* \cdot \nabla^*\right)\mathbf{v}^* = \rho g \mathbf{g}^* - \frac{p_o - p_\infty}{L}\nabla^* p^* + \frac{\mu U_\infty}{L^2}\nabla^{*2}\mathbf{v}^*$$

$$(3.2)$$

All terms now have dimensions $\{ML^{-2}T^{-2}\}$, and so to non-dimensionalize Eq. (3.2) we must multiply each term by a constant $\frac{L}{\rho U_\infty^2}\{M^{-1}L^2T^2\}$ in order to cancel out all physical dimensions:

$$\left[\frac{fL}{U_\infty}\right]\frac{\partial \mathbf{v}^*}{\partial t^*} + \left(\mathbf{v}^* \cdot \nabla^*\right)\mathbf{v}^* = \left[\frac{gL}{U_\infty^2}\right]\mathbf{g}^* - \left[\frac{p_o - p_\infty}{\rho U_\infty^2}\right]\nabla^* p^* + \left[\frac{\mu}{\rho U_\infty L}\right]\nabla^{*2}\mathbf{v}^*.$$

$$(3.3)$$

The following describes each key dimensionless group, and at this stage, a first impression of their physical interpretation:

$\left[\frac{fL}{U_\infty}\right]$ is the Strouhal number describing the level of unsteadiness or temporal variation in the fluid flow, where $\text{St} = \frac{fL}{U_\infty}$;

$\left[\frac{gL}{U_\infty^2}\right]$ is the inverse of the Froude number squared, where the Froude number $\text{Fr} = \frac{U_\infty}{\sqrt{gL}}$ describes the influence of gravity such as free-surface effects on the system;

$\left[\frac{p_o-p_\infty}{\rho U_\infty^2}\right]$ is the Euler number capturing the relationship between pressure and inertial forcing in the flow, where $\text{Eu} = \frac{p_o-p_\infty}{\rho U_\infty^2}$; and

$\left[\frac{\mu}{\rho U_\infty L}\right]$ is the inverse of the Reynolds number, where the Reynolds number $\text{Re} = \frac{\rho U_\infty L}{\mu}$ describes the relative strength of inertial versus viscous forces, and along with that, also an indication of the laminar, transitional, or fully turbulent state of the flow itself.

Finally, we can rewrite Eq. (3.3) in a more compact and insightful form as follows:

$$\text{St}\frac{\partial \mathbf{v}^*}{\partial t^*} + \left(\mathbf{v}^* \cdot \nabla^*\right)\mathbf{v}^* = \frac{1}{\text{Fr}^2}\mathbf{g}^* - \text{Eu}\nabla^* p^* + \frac{1}{\text{Re}}\nabla^{*2}\mathbf{v}^*. \qquad (3.4)$$

> **Tips**
>
> - Non-dimensionalization concerns only dimensions of the equation and we can use any value for scaling parameters; whereas
> - Normalization is more restrictive—here we must choose scaling parameters that are appropriate for the given problem.

In order for all non-dimensional variables (t^*, $\mathbf{v}^*$, p^*, etc.) to be of magnitude unity (i.e., $t^* \sim 1$, $\mathbf{v}^* \sim 1$, $p^* \sim 1$) we must pick problem-specific scaling parameters. If we properly normalize the Navier–Stokes equation, we can judge the relative importance of various terms (and their physical meaning) by comparing the relative magnitudes of St, Fr, Eu, Re, etc. In Fig. 3.1, we present a simple example where in each specific region we can identify when key dimensionless groups are zero, and hence critical terms of the Navier–Stokes equation vanish. Recall, for instance, that an irrotational flow (i.e., $\boldsymbol{\omega} = \nabla \times \mathbf{v} = 0$) implies an inviscid flow (i.e., $Re \Rightarrow \infty$), but the opposite is not always true, and an inviscid flow can in fact be rotational.

3.2 Stokes Flow

Let us begin by considering the very smallest of scales ($Re \Rightarrow 0$). In Stokes flow (Stokes 1842), also known as creeping flow or low-Reynolds-number flow, inertial forces (convective terms) are found to be negligibly small when compared to pressure and viscous forcing. To satisfy the Stokes-flow condition we require the following:

- Quasi-steady flow with negligibly small Strouhal number ($St \ll 1$);
- A low Reynolds number ($Re \ll 1$) such that convective terms are $\mathcal{O}(1)$; and
- Body-force effects, such as those caused by a free surface, are negligible where $Fr \gg 1$.

Starting from the normalized Navier–Stokes equation as presented above, we can cancel several terms:

$$St\frac{\partial \mathbf{v}^*}{\partial t^*} + \left(\mathbf{v}^* \cdot \nabla^*\right)\mathbf{v} = \frac{1}{Fr^2}\mathbf{g}^* - Eu\nabla^* p^* + \frac{1}{Re}\nabla^2 \mathbf{v}^*, \tag{3.5}$$

and when simplifying we obtain a simple yet elegant expression for Stokes flow where:

$$\nabla p \cong \mu \nabla^2 \mathbf{v}. \tag{3.6}$$

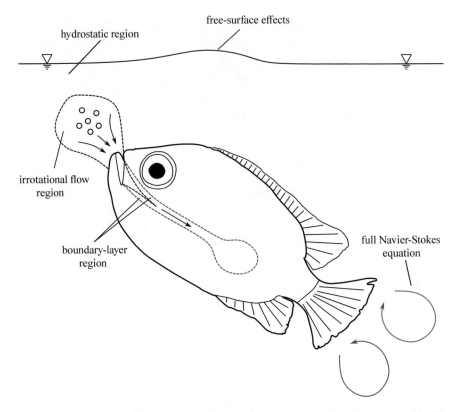

Fig. 3.1 Filter feeding near the water surface provides a cute example where we can identify distinct flow-field behavior where key dimensionless groups tend to zero, and the normalized Navier–Stokes equation correspondingly simplifies. Some regions of the fluid are clearly hydrostatic while others behave in an inviscid manner, i.e. viscous effects are negligible and only inertial and pressure terms remain (depicted as irrotational region). Any distortion of the water (free) surface will trigger forcing described by the Froude number, and in the unsteady wake of the fish's tail, both Strouhal and Reynolds numbers will be key (non-negligible) parameters. Finally, the developing flow passing food into the fish's stomach, along with the thin layer of fluid passing over its body, can be simplified by the boundary-layer approximation, as seen later in this chapter

It is worth mentioning that since a Stokes flow behaves in a quasi-steady manner (i.e., St≪1), and the flow has zero inertia, then in turn the hysteresis will also be zero. In other words, if a cell were to swim along an arbitrary path from A to B, it could move backwards from B to A using exactly the same oscillatory motion but in reverse. We will discuss some examples of cellular motility shortly.

Tips

The following outlines some key assumptions necessary for the simplification to a Stokes flow:

- Quasi-steady flow such that $\text{St}\frac{\partial \mathbf{v}^*}{\partial t^*} = 0$;
- Negligibly small inertial forces $\rightarrow (\mathbf{v}^* \cdot \nabla^*)\,\mathbf{v} = 0$; and
- Body forces such as free-surface effects are negligible (e.g. $\text{Fr} \gg 1$), where for instance $\frac{1}{\text{Fr}^2}\mathbf{g}^* = 0$.

For the normalized form of the Navier–Stokes equation (Eq. (3.5)), the balance between the pressure and viscous terms in Stokes flow is only possible when the Euler number (Eu) is on the same order as the inverse of the Reynolds number $\left(\frac{1}{\text{Re}}\right)$ such that:

$$\frac{p_o - p_\infty}{\rho U_\infty^2} \sim \frac{\mu}{\rho U_\infty L}. \tag{3.7}$$

We now observe an unusual shift from our experience in the inertial world (i.e., high Re) where the pressure now scales linearly with the velocity (rather than with the velocity squared):

$$p_o - p_\infty \sim \frac{\mu U_\infty}{L}. \tag{3.8}$$

In fact, this curious behavior, where pressure scales linearly with velocity, shares some commonalities with our experience relating friction to velocity in solid mechanics, i.e., we know that friction often scales linearly with interface velocity.

Tips

Some further points to consider when relating our first experiences with Stokes flow to biological systems:

- This is a dramatic shift away from inertially dominated flows where $p_o - p_\infty \sim \rho U_\infty^2$ (recall Bernoulli equation);
- Fluid density has dropped out entirely for Stokes flow;
- Examples in nature include *flagellates* and *cilia* (flagellar movement);
- Two types of propulsive methods in a Stokes flow include: (1) Prokaryotes (Archaea & Bacteria $\text{Re} = 10^{-6}$) and (2) Eukaryotes (Spermatozoa $\text{Re} = 10^{-3}$); and
- Stokes flow also describes the passive motion of small droplets (e.g. aerosols) or drifting cells.

For the passive motion of small bodies, we can define yet another key dimensionless group known as the Stokes number (Stk):

$$Stk = \frac{\tau}{\tau_f},$$ (3.9)

where τ defines the relaxation time of the body (time it takes for body to accelerate to new external conditions due to drag), and τ_f represents the characteristic time scale of the flow itself, e.g., $\tau_f = \frac{L}{U_\infty}$. We will tackle the Stokes drag on a generic sphere shortly.

For the case of a drifting body (e.g., a passive cell), a general rule of thumb is that for Stk < 0.1, the deviation of the body's trajectory relative to the fluid path is on the order of less than 1%. This deviation, however, becomes much more pronounced as Stk increases, as shown qualitatively in Fig. 3.2. Let us now contrast the propulsive modes between low and high Re (Stokes flow versus inertial conditions at high Re), as shown in Fig. 3.3. Beginning with high Re, we can clearly relate the motion of a flapping wing or oscillating fin to propulsion as shown in Fig. 3.4.

For this high Reynolds number (inertial) flow, the animal can glide during and between oscillatory input of momentum. In this case, the expression for drag (D) is generally scaled as:

$$D = C_D \frac{1}{2} \rho U_\infty{}^2 A,$$ (3.10)

where C_D is the drag coefficient characteristic of the body shape, U_∞ is the body velocity, and A represents the characteristic area, such as the frontal area.

Fig. 3.2 Deviation of passive body from fluid path when the Stokes number is sufficiently high (e.g., Stk > 0.1). Large deviations are expected as the Stokes number increases, i.e., through a larger deviation between body and fluid densities (e.g., passive motion in air rather than water)

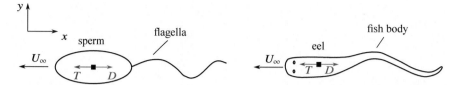

Fig. 3.3 Example of low Reynolds (Stokes) flow for cellular motility on left in contrast to high-Reynolds-number propulsion of an inertial swimmer on the right

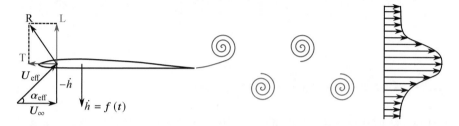

Fig. 3.4 Motion of a flapping wing or oscillating fin generating propulsive force

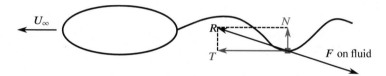

Fig. 3.5 Tangential force of each flagellum section on the surrounding fluid

At low Re (Stokes flow), on the other hand, there is no gliding present, i.e., when flagellar motion stops, the body must stop instantaneously as well. As noted earlier through Eq. (3.8), the expression for the reaction force (drag) associated with flagellar motion is therefore:

$$D = C_D \mu U_\infty L, \tag{3.11}$$

where as before, we note that the pressure (and hence overall force) now scales linearly with velocity U_∞ instead. For the motion of a flagellar swimmer, as shown in Fig. 3.5, one can imagine how each segment of the flagellum generates an equal and opposite reaction force (thrust T) able to overcome the body's drag.

3.2.1 Exercise for Stokes' Drag on a Sphere

Let us examine Stokes' solution for the flow past an immersed sphere, where a sphere represents a first simple yet helpful abstraction of many active (and passive) swimmers, cells, etc. To start let us consider some of the key conditions for this problem:

1. Consider a creeping flow with velocity U_∞ past a solid sphere of radius a, as shown in Fig. 3.6.

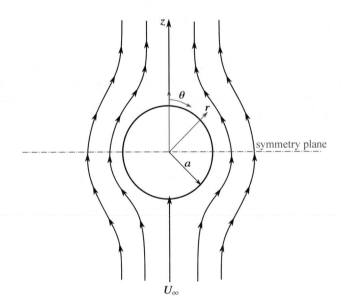

Fig. 3.6 Considering the flow past a sphere moving with velocity U_∞. Note fore-aft symmetry plane showing identical streamline curvature

2. To solve for the flowfield $\mathbf{v} = (v_r, v_\theta, v_\phi)$ throughout, as well as the forces experienced by the body itself, we must solve both the continuity and Navier–Stokes equations, where we assume no swirl, such that $v_\phi = 0$, and through symmetry that $\dfrac{\partial}{\partial \phi} = 0$.

We assume an incompressible flow such that continuity:

$$\nabla \cdot \mathbf{v} = 0 \tag{3.12}$$

is satisfied. For the Navier–Stokes equation, inertia is neglected for a Stokes' flow (Re < 1) such that we obtain

$$\nabla p \approx \mu \nabla^2 \mathbf{v}, \tag{3.13}$$

where the pressure gradient ∇p is counteracted through viscous resistance.

We must now delve into the world of spherical coordinates where continuity is written:

$$\frac{1}{r^2} \frac{\partial}{\partial r}(r^2 v_r) + \frac{1}{r} \frac{\partial}{\partial \theta}(v_\theta \sin \theta) = 0. \tag{3.14}$$

In order to solve this expression, we define a stream function ψ that satisfies continuity identically. We may be familiar with ψ in Cartesian coordinates, where

$$\frac{du}{dx} + \frac{dv}{dy} = 0, \tag{3.15}$$

which is satisfied if

$$\frac{\partial}{\partial x}\left(\frac{\partial \psi}{\partial y}\right) + \frac{\partial}{\partial y}\left(-\frac{\partial \psi}{\partial x}\right) = 0. \tag{3.16}$$

It follows therefore that

$$u = \frac{\partial \psi}{\partial y}, \, v = -\frac{\partial \psi}{\partial x}, \tag{3.17}$$

and

$$\mathbf{v} = \frac{\partial \psi}{\partial y}\hat{i} - \frac{\partial \psi}{\partial x}\hat{j}, \tag{3.18}$$

such that lines of constant ψ are effectively streamlines (and hence do not cross).

In a similar manner, we can define our velocity field in terms of ψ in spherical coordinates such that:

$$v_r = \frac{1}{r^2 \sin\theta}\frac{\partial\psi}{\partial\theta}, \qquad v_\theta = -\frac{1}{r\sin\theta}\frac{\partial\psi}{\partial r}, \tag{3.19}$$

which thus satisfies the continuity equation (Eq. (3.12)). With these definitions using ψ, we can rewrite Eq. (3.13):

$$\nabla p \approx \mu\nabla^2\mathbf{v} \quad\Rightarrow\quad \left(\frac{\partial^2}{\partial r^2} + \frac{\sin\theta}{r^2}\frac{\partial}{\partial\theta}\left(\frac{1}{\sin\theta}\frac{\partial}{\partial\theta}\right)\right)^2 \psi = 0. \tag{3.20}$$

This expression is a second-order PDE (whose derivation is beyond the scope of this exercise) to which we can apply the following boundary conditions:

1. At the sphere surface, $r = a$: $v_r = 0$, $v_\theta = 0$ (no-slip).
2. At infinity, $r \to \infty$: v_r and v_θ are functions of freestream U_∞, where $v_r = U_\infty \cos\theta$, and $v_\theta = -U_\infty \sin\theta$.

We now write ψ in terms of r and θ for $r \to \infty$:

$$v_\theta = -U_\infty \sin\theta = -\frac{1}{r\sin\theta}\frac{\partial\psi}{\partial r}, \tag{3.21}$$

and integrate

$$\psi = \int r U_\infty \sin^2\theta\, dr = \frac{U_\infty}{2} r^2 \sin^2\theta. \tag{3.22}$$

Using the two boundary conditions, we can now solve

$$\psi = \frac{U_\infty}{2}\left[r^2 + \frac{a^3}{2r} - \frac{3ar}{2}\right]\sin^2\theta, \tag{3.23}$$

where r^2 represents the uniform flow, $\frac{a^3}{2r}$ a doublet, and $\frac{3ar}{2}$ a viscous correction (this last term disappears in a potential flow).

We, therefore, obtain the velocity field $\mathbf{v} = (v_r, v_\theta, 0)$ around the sphere (in creeping flow):

$$v_r = \frac{1}{r^2\sin\theta}\frac{\partial\psi}{\partial\theta} = U_\infty \cos\theta\left[1 + \frac{a^3}{2r^3} - \frac{3a}{2r}\right], \tag{3.24}$$

$$v_\theta = -\frac{1}{r\sin\theta}\frac{\partial\psi}{\partial r} = -U_\infty \sin\theta\left[1 + \frac{a^3}{4r^3} - \frac{3a}{4r}\right]. \tag{3.25}$$

We can now draw some conclusions from **v** and calculate the forces on the sphere. But first, let us yet again observe how ψ and **v** components are independent of fluid viscosity in a Stokes flow. Furthermore, unlike inertial (high-Re) flows, the local velocity is slowed down everywhere near the body, i.e., the sphere's influence on the fluid velocity is felt even at very large distances away in the absence of inertia. For instance, even at $r = 10a$, the velocity field is still only 90% of U_∞.

When examining the flow topology (streamlines) around the sphere in question, we note how, unlike for inertial problems, there is perfect fore-and-aft symmetry, as shown in Fig. 3.6. If asymmetric, the pressure distribution will impose a non-zero force on the sphere surface. We can extract the pressure term, which acts toward the center of the sphere from all directions, by examining the r-component of the Stokes equation (3.13) using the following identity:

$$\nabla \times (\nabla \times \mathbf{v}) = \nabla(\nabla \cdot \mathbf{v}) - \nabla^2 \mathbf{v} \tag{3.26}$$

we obtain

$$\frac{\partial p}{\partial r} = -\mu \left[\nabla \times (\nabla \times \mathbf{v})\hat{r} \right], \tag{3.27}$$

where the term in square brackets is the curl of the vorticity field:

$$\boldsymbol{\omega} = \nabla \times \mathbf{v} = \frac{1}{r\sin\theta} \left(\frac{\partial}{\partial\theta}(v_\phi \sin\theta) - \frac{\partial v_\theta}{\partial\phi} \right) \hat{r} + \frac{1}{r} \left(\frac{1}{\sin\theta} \frac{\partial v_r}{\partial\phi} - \frac{\partial}{\partial r}(r v_\phi) \right) \hat{\theta}$$
$$+ \frac{1}{r} \left(\frac{\partial}{\partial r}(r v_\theta) - \frac{\partial v_r}{\partial\theta} \right) \hat{\phi}. \tag{3.28}$$

We can simplify this expression for the vorticity field knowing that $v_\phi = 0$ and $\frac{\partial}{\partial\phi} = 0$:

$$\boldsymbol{\omega} = \frac{1}{r} \left(\frac{\partial}{\partial r}(r v_\theta) - \frac{\partial v_r}{\partial\theta} \right) \hat{\phi}$$
$$= \frac{1}{r} \left[-U_\infty \sin\theta (1 + \frac{a^3}{2r^3}) + U_\infty \sin\theta (1 + \frac{a^3}{2r^3} - \frac{3a}{2r}) \right] \hat{\phi} \tag{3.29}$$
$$\boldsymbol{\omega} = -\frac{3aU_\infty}{2r^2} \sin\theta \hat{\phi}.$$

We now return to the expression in the square brackets of Eq. (3.27):

$$\nabla \times \boldsymbol{\omega} = \nabla \times (\nabla \times \mathbf{v}) = \frac{1}{r \sin \theta} \left(-\frac{3aU_\infty}{2r^2} \cdot 2 \sin \theta \cos \theta \right) \hat{r} - \frac{1}{r} \left(\frac{3aU_\infty}{2r^2} \sin \theta \right) \hat{\theta}$$

$$= \frac{3aU_\infty}{r^3} \cos \theta \hat{r} - \frac{3aU_\infty}{2r^3} \sin \theta \hat{\theta}.$$

(3.30)

Since we only need the first term to obtain the radial pressure distribution on the sphere surface, we reduce Eq. (3.27) to

$$\frac{\partial p}{\partial r} = \frac{3\mu U_\infty a}{r^3} \cos \theta$$

(3.31)

and integrate with respect to radius r, from r to ∞:

$$p(r) = \int_r^\infty \frac{3\mu U_\infty a}{r^3} \cos \theta \, dr = 3\mu U_\infty a \cos \theta \left[-\frac{1}{2r^2} \right]_r^\infty$$

(3.32)

$$p(r) = p_\infty - \frac{3}{2} \frac{\mu U_\infty a}{r^2} \cos \theta \quad .$$

We now observe that p is antisymmetric such that the pressure is positive on the front face of the sphere, whereas the pressure is negative on the backside. Despite appearances (streamlines), this asymmetry in pressure generates drag (pressure force) on the sphere.

In a similar manner, we must also consider the shear stress on the sphere surface as it too will contribute to the total drag force. Since we only care about shear tangential to the sphere surface, the surface shear stress can be defined as

$$\tau_{r\theta} = \mu \epsilon_{r\theta},$$

(3.33)

where we recall that ϵ is the strain rate. We can now expand in spherical coordinates:

$$\tau_{r\theta} = \mu \left[\frac{1}{r} \frac{\partial v_r}{\partial \theta} + \frac{\partial v_\theta}{\partial r} - \frac{v_\theta}{r} \right]$$

(3.34)

and substitute in Eqs. (3.24) and (3.25) such that:

$$\tau_{r\theta} = -\frac{3\mu a^3 U_\infty}{2r^4} \sin \theta,$$

(3.35)

or when calculated directly at the surface $r = a$:

$$\tau_{r\theta} - \frac{3\mu U_\infty}{2a} \sin \theta.$$

(3.36)

Finally, the total drag on the sphere will equal the sum of both the pressure and shear forces over the entire surface of the sphere:

$$D = -\int_0^\pi \tau_{r\theta}\big|_{r=a} \sin\theta \, dA - \int_0^\pi p\big|_{r=a} \cos\theta \, dA, \tag{3.37}$$

where $dA = 2\pi a^2 \sin\theta \, d\theta$. Note that the factor 2 accounts for the limits from 0 to π instead of to 2π, such that the drag on a sphere in Stokes flow (Re $\ll$ 1) can be written as follows:

$$D = 4\pi \mu U_\infty a + 2\pi \mu U_\infty a = 6\pi \mu U_\infty a, \tag{3.38}$$

where it should be noted that the first term $(4\pi \mu U_\infty a)$ is the contribution due to shear and accounts for two thirds of the total drag, whereas the second term $(2\pi \mu U_\infty a)$—one third of the total drag—represents the pressure difference between front and back faces of the sphere. We can also define the total sphere drag coefficient as is commonly presented in the literature (as scaled with the high-Re definition and hence proportional to U_∞^2 rather than U_∞):

$$C_D = \frac{D}{\frac{1}{2}\rho U_\infty^2 \pi A} = \frac{6\pi \mu U_\infty a}{\frac{1}{2}\rho U_\infty^2 \pi a^2} = \frac{24}{Re}. \tag{3.39}$$

Thus, what this detailed exercise has shown us is that, when shrunk down to the world of a cell, or perhaps drifting in the atmosphere like a plumed seed, the hydrodynamic (or aerodynamic) drag is absolutely enormous in contrast to our experience and intuition at human scales. If you can imagine swimming in a vat of honey, then perhaps you are ready to accept the fact that inertia is truly of little performance under these Stokesian conditions!

3.3 Boundary-Layer Approximation

Beginning from our normalized Navier–Stokes equation (Eq. (3.4)) and now accounting for inertial effects, let us consider steady flow past a slender, streamlined body, as shown in Fig. 3.7:

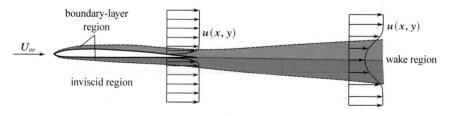

Fig. 3.7 Flow past a streamlined body depicting both viscous and inviscid regions

$$\left(\mathbf{v}^* \cdot \nabla^*\right) \mathbf{v}^* = -\mathrm{Eu}\nabla^* p^* + \frac{1}{\mathrm{Re}}\nabla^{*2}\mathbf{v}^*. \tag{3.40}$$

Away from the body surface and the wake region, when $Re \gg 1$, viscous forces are negligible when compared to inertial and/or pressure forces. As such these regions (outside the boundary layers) are referred to as *inviscid* and can be treated with the Euler equation (e.g., Bernoulli). Furthermore, vorticity is negligibly small outside of the thin boundary-layer and wake regions. Note that irrotational flow implies an inviscid region but that the converse is not necessarily true.

Let us now take a closer look at the region near the body surface, or in its wake, where vorticity is generated and subsequently convected downstream. Here, solving the Navier–Stokes equation directly, whether in dimensional or normalized form, is impractical in most cases. Besides we are looking to derive physical insight into the flow's physics, for which the Navier–Stokes equation is generally too complex. To treat these relatively thin viscous regions, Prandtl (1904) first introduced the concept of the *boundary-layer approximation*, which is described as follows. When $Re \gg 1$, vorticity (generated at the wall through the no-slip condition) is swept downstream in a thin layer ($\delta(x)$) such that:

$$\delta \ll x, \quad v \ll u, \quad \frac{\partial u}{\partial x} \ll \frac{\partial u}{\partial y} \quad \text{and} \quad \frac{\partial v}{\partial x} \ll \frac{\partial v}{\partial y}. \tag{3.41}$$

Physically, the above approximations imply that the rate of convection parallel to the body surface is much larger than the rate of transverse (viscous) diffusion, be it laminar or turbulent. This competition between convection and diffusion is shown in Fig. 3.8.

More generally, we can use the approximation when the viscous region δ is sufficiently thin, as defined by

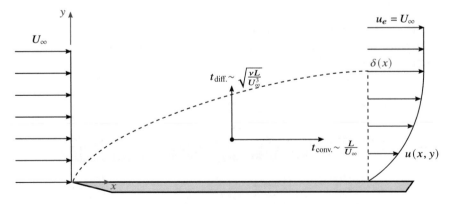

Fig. 3.8 Competition between convection and diffusion in a generic boundary layer

$$\sqrt{\frac{vL}{U_\infty^3}} \ll \frac{L}{U_\infty} \text{ or } \sqrt{\text{Re}_L} \gg 1, \tag{3.42}$$

where L represents the characteristic length scale of the problem, such as the length of the body or the wing chord.

The thin boundary-layer assumption should therefore hold if Re_L is sufficiently large, but the approximation's validity is also very much geometry/shape specific. Furthermore, one should consider that the approximation has various levels of accuracy and that even at lower Reynolds numbers the basic principles still qualitatively hold.

Despite its name, the boundary-layer approximation can just as well be used to treat free shear layers present in wakes, jets, and mixing layers so long as $\text{Re}_L \gg 1$ and $\delta/x \ll 1$. Figure 3.9 shows where the approximation holds for a bluff body (generic cylinder wake). In Fig. 3.10 we see analogous examples for generic jets and mixing layers as well.

Tips

Within a boundary-layer region there must be:

- Non-negligible viscous forces;
- Finite vorticity; and
- Solid-walls or body forces to generate vorticity through shear.

Now let us proceed in Cartesian coordinates and normalize the two-dimensional continuity and Navier–Stokes equations using the following approximations derived from boundary-layer theory:

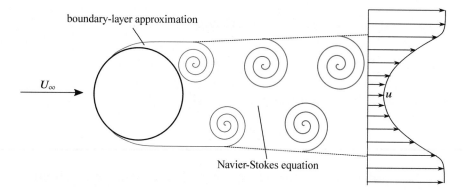

Fig. 3.9 Boundary-layer approximation holds in the shear layer and far wake of a generic bluff body like a circular cylinder

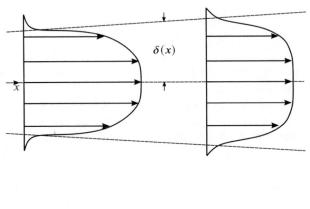

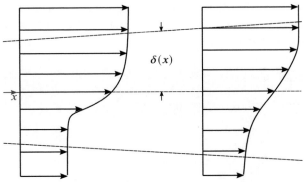

Fig. 3.10 Boundary-layer approximation for both canonical jets (top) and mixing layers (bottom)

$$u \backsim U_\infty, \ p - p_\infty \backsim \rho U_\infty^2, \ \frac{\partial}{\partial x} \backsim \frac{1}{L}, \ \text{and} \ \frac{\partial}{\partial y} \backsim \frac{1}{\delta}. \tag{3.43}$$

When substituting into the two-dimensional form of the continuity equation

$$\frac{\partial u}{\partial x} + \frac{\partial v}{\partial y} = 0, \tag{3.44}$$

we obtain the following relationship:

$$\frac{U_\infty}{L} \backsim \frac{V}{\delta} \ \text{or} \ V \backsim \frac{U_\infty \delta}{L}, \tag{3.45}$$

and since $\frac{\delta}{L} \ll 1$ in the thin boundary-layer region, we obtain very generally that

$$v \ll u. \tag{3.46}$$

We now define the following non-dimensional normalized variables:

$$x^* = \frac{X}{L}, \quad y^* = \frac{Y}{\delta}, \quad u^* = \frac{u}{U_\infty}, \quad v^* = \frac{VL}{U_\infty \delta} \quad \text{and} \quad p^* = \frac{p - p\infty}{\rho U_\infty^2}. \tag{3.47}$$

Beginning with the y-component of momentum from the Navier–Stokes equation

$$u\frac{\partial v}{\partial x} + v\frac{\partial v}{\partial y} = -\frac{1}{\rho}\frac{\partial p}{\partial y} + v\frac{\partial^2 v}{\partial x^2} + v\frac{\partial^2 x}{\partial y^2}, \tag{3.48}$$

we can substitute in the non-dimensionalizations from Eq. (3.47) and then multiply by $\frac{L^2}{U_\infty^2 \delta}$ to obtain the following:

$$u^*\frac{\partial v^*}{\partial x^*} + v^*\frac{\partial v^*}{\partial y^*} = -\left(\frac{L}{\delta}\right)^2 \frac{\partial p^*}{\partial y^*} + \left(\frac{v}{U_\infty L}\right)\frac{\partial^2 v^*}{\partial x^{*2}} + \left(\frac{v}{U_\infty L}\right)\left(\frac{L}{\delta}\right)^2 \frac{\partial^2 x^*}{\partial y^{*2}}. \tag{3.49}$$

Based on the inequalities in Eq. (3.41), the above equation can be simplified. The first two terms are negligible within the boundary layer where $v \ll u$ and $\frac{\partial v}{\partial x} \ll \frac{\partial v}{\partial y}$. For Re $\gg 1$, $\frac{v}{U_\infty L} \rightsquigarrow 0$, and the term containing the pressure gradient is dominant (where $L \gg \delta$). Thus, ultimately, Eq. (3.49) for y-momentum reduces to the following result:

$$\frac{\partial p^*}{\partial y^*} \approx 0 \tag{3.50}$$

and, in dimensional form, simply

$$\frac{\partial p}{\partial y} \approx 0. \tag{3.51}$$

Tips

Some interesting outcomes from the boundary-layer approximation are as follows:

- Streamlines are assumed to have negligible curvature in thin boundary layers;
- With curvature, centripetal acceleration requires balancing the pressure gradient along the radius of curvature; and
- Pressure in the outer (inviscid) region now can be obtained in relation to wall pressure or *vice versa*.

Following along the same rationale, the x-component of momentum from the Navier–Stokes equation

$$u\frac{\partial u}{\partial x} + v\frac{\partial u}{\partial y} = -\frac{1}{\rho}\frac{dp}{dx} + v\frac{\partial^2 u}{\partial x^2} + v\frac{\partial^2 u}{\partial y^2}, \tag{3.52}$$

where it is now known that $p \neq f(y)$, can be manipulated as follows. As before, we substitute in the non-dimensionalizations (Eq. (3.47)) and multiply throughout by $\frac{L^2}{U_\infty^2}$, thus obtaining

$$u^*\frac{\partial u^*}{\partial x^*} + v^*\frac{\partial u^*}{\partial y^*} = -\frac{dp^*}{dx^*} + \left(\frac{v}{U_\infty L}\right)\frac{\partial^2 u^*}{\partial x^{*2}} + \left(\frac{v}{U_\infty L}\right)\left(\frac{L}{\delta}\right)^2\frac{\partial^2 u^*}{\partial y^{*2}}. \tag{3.53}$$

This time the velocity gradients are non-negligible and the coefficient of the final term is of order unity, such that $\frac{\delta}{L} \sim \frac{1}{\sqrt{Re_L}}$ when $Re_L \gg 1$. Inside the boundary layer, the x-momentum equation becomes

$$u\frac{\partial u}{\partial x} + v\frac{\partial u}{\partial y} = -\frac{1}{\rho}\frac{dp}{dx} + v\frac{\partial^2 u}{\partial y^2}. \tag{3.54}$$

However, since $p \neq f(y)$, the pressure gradient can be related to that within the inviscid (outer) region through the Bernoulli equation:

$$\frac{p}{\rho} + \frac{U_\infty^2}{2} = \text{const.} \tag{3.55}$$

Taking the derivative along the x-direction, we can equate the pressure gradient to the outer velocity U_e through

$$\frac{1}{\rho}\frac{dp}{dx} = -U_e\frac{dU_e}{dx}. \tag{3.56}$$

We then obtain the second and final equation for the boundary-layer approximation:

$$u\frac{\partial u}{\partial x} + v\frac{\partial u}{\partial y} = U_e\frac{dU_e}{dx} + v\frac{\partial^2 u}{\partial y^2}, \tag{3.57}$$

where $U_e = f(x)$. Note that Eqs. (3.51) and (3.57) represent the governing equations for the boundary-layer approximation.

Tips

Unlike for the Navier–Stokes equation (elliptic PDE), the x-momentum boundary-layer equation is parabolic → information in flow does not pass upstream!

3.3.1 Exercise on Jellyfish Propulsion and Self-Similar Jets

Consider a jellyfish swimming from right-to-left as shown in Fig. 3.11. In principle, the deforming body propels itself via periodic bell contractions thus producing a jet of water downstream. Each contraction produces a distinct vortex ring, perhaps not any different than those produced in our aorta, as was shown and discussed in Fig. 2.13. These intensely vortical structures then convect downstream away from the jellyfish while subsequently diffusing into the mean flow.

Suppose we want to first determine the mean (time-averaged) velocity profile of the jet downstream (as illustrated in Fig. 3.12), and in turn characterize the mean thrust generated by the jellyfish motion. In other words, we will first blur out distinct vortical pulses and average out their effect over time and space.

As always, we start with the continuity and Navier–Stokes equations, and by using reasonable first assumptions, we can simplify sufficiently (i.e., let us take the above boundary-layer approximations into consideration).

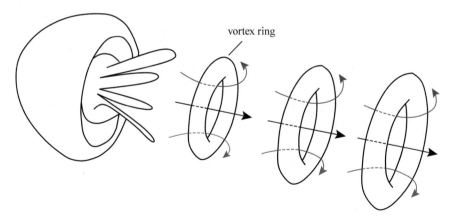

Fig. 3.11 Jellyfish propelling itself from right-to-left by shedding periodic vortex rings into its wake

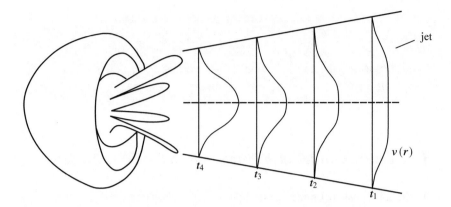

Fig. 3.12 A time-averaged and thus simplified (mean) representation of the jet flow behind the jellyfish. Here we are keeping things very simple by blurring the individual pulses in the wake. However, we will revisit this interesting and critical unsteady problem in a latter section

The following are some reasonable assumptions for our problem:

- $v_\theta = 0 \rightarrow$ axisymmetric flow (also $\partial/\partial\theta = 0$).
- $v_r \ll v_z \rightarrow$ velocity profile dominated by body motion in the z-direction.

Similarly, $Re_z \gg 1 \Leftrightarrow \frac{U_z}{v} \gg 1$, $\delta \ll z$, such that:

- $\dfrac{\partial v_z}{\partial z} \ll \dfrac{\partial v_z}{\partial r} \rightarrow$ gradient much stronger in the radial direction; item $\dfrac{\partial v_r}{\partial z} \ll \dfrac{\partial v_r}{\partial r}$
  $\rightarrow$ same as above and also recall that v_r is small to begin with.
- $\partial/\partial t = 0 \rightarrow$ steady flow.

Let us now consider the Navier–Stokes equation starting from vector form:

$$\rho \frac{D\mathbf{v}}{Dt} = -\nabla p + \mu \nabla^2 \mathbf{v} + \rho \mathbf{g}. \tag{3.58}$$

Expanding to the three directional momentum equations, the r-component of momentum reduces to

$$\rho \left(\cancel{\frac{\partial v_r}{\partial t}} + v_r \cancel{\frac{\partial v_r}{\partial r}} + \cancel{\frac{v_\theta}{r} \frac{\partial v_r}{\partial \theta}} + v_z \cancel{\frac{\partial v_r}{\partial z}} - \cancel{\frac{v_\theta^2}{r}} \right)$$

$$= -\frac{\partial p}{\partial r} + \cancel{\rho g_r} + \mu \left[\frac{1}{r} \frac{\partial}{\partial r} \left(r \cancel{\frac{\partial v_r}{\partial r}} \right) - \cancel{\frac{v_r}{r^2}} + \cancel{\frac{1}{r^2} \frac{\partial^2 v_r}{\partial \theta^2}} - \cancel{\frac{2}{r^2} \frac{\partial v_\theta}{\partial \theta}} + \cancel{\frac{\partial^2 v_r}{\partial z^2}} \right],$$

$$\tag{3.59}$$

where, applying the above assumptions, most terms are either zero or negligibly small compared to others. Thus, we obtain simply that

$$\frac{\partial p}{\partial r} \simeq 0 \tag{3.60}$$

and hence the pressure field reduces to $p = f(z)$ only.

In a similar manner the z-component of momentum becomes

$$\rho \left(\frac{\cancel{\partial v_z}}{\cancel{\partial t}} + v_r \frac{\partial v_z}{\partial r} + \frac{v_\theta \cancel{\partial v_z}}{\cancel{r} \, \partial \theta} + v_z \frac{\partial v_z}{\partial z} \right)$$
$$= -\frac{\cancel{\partial p}}{\cancel{\partial z}} + \rho \cancel{g_z} + \mu \left[\frac{1}{r} \frac{\partial}{\partial r} \left(r \frac{\partial v_z}{\partial r} \right) + \frac{1}{r^2} \frac{\cancel{\partial^2 v_z}}{\cancel{\partial \theta^2}} + \frac{\cancel{\partial^2 v_r}}{\cancel{\partial z^2}} \right], \tag{3.61}$$

in turn reducing simply to the following:

$$v_r \frac{\partial v_z}{\partial r} + v_z \frac{\partial v_z}{\partial z} = v \left[\frac{1}{r} \frac{\partial}{\partial r} \left(r \frac{\partial v_z}{\partial r} \right) \right]. \tag{3.62}$$

Note that the pressure gradient was found to equal zero using Eq. (3.56), where the jellyfish is assumed to be moving through a quiescent fluid and $U_e = 0$.

Repeating the same procedure for the θ-momentum, you will find each term is zero when $v_\theta = 0$.

Now, with Eqs. (3.60) and (3.62), we can try to solve for the velocity field behind the jellyfish using the concept of *self-similarity*. In fluid dynamics, a self-similar solution applies to flows for which some scaling factor or factors exist that scale the flow along an axis of its motion, i.e., usually the velocity profile looks more-or-less the same at any scale.

For instance, for a simple flat-plate boundary-layer example, as shown in Fig. 3.13, we can find scaling factors a and b such that:

$$u(x + \Delta x, y/a) = bu(x, y). \tag{3.63}$$

One famous self-similar solution is known as the Blasius solution (Blasius 1908) for a boundary layer on an infinitely long flat plate. However, this solution can just as well be used for other flows that follow the boundary-layer approximation, such as plane laminar jets and wakes.

For our jellyfish example, self-similarity can be applied to describe the jet flow provided an appropriately large Reynolds number exists to satisfy the boundary-layer approximation, i.e., $\frac{\delta}{z} \ll 1$, as shown in Fig. 3.14. However, we lose similarity if the flow transitions to turbulence at even higher Reynolds numbers, as evidenced by the so-called *Kelvin–Helmholtz* instabilities that form on the edge of the jet, which in turn roll up and strongly affect the mean velocity profiles.

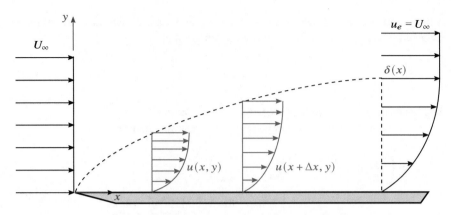

Fig. 3.13 Scaling of two flat-plate boundary-layer profiles with one another

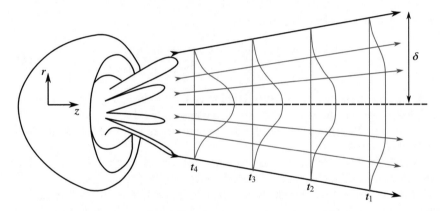

Fig. 3.14 Velocity profiles are similar at each z-position with parallel streamlines that maintain separation with one another

Returning now to the Blasius solution, and applying our boundary conditions:

1. axisymmetric ($v_r = 0$, $\dfrac{\partial v_z}{\partial r} = 0$ at $r = 0$); and
2. quiescent ambient fluid ($v_z = 0$ at $r = \infty$),

we can solve the boundary-layer equation and write our solution in terms of the momentum flux (J) across a cross-section of the jet:

$$J = \rho \int_{-\infty}^{\infty} v_z^2 \, dr. \tag{3.64}$$

The resultant velocity profile takes on the form:

$$v(z, r) = v_{max} \operatorname{sech}^2 \left[0.2752 \left(\frac{J\rho}{\mu^2 z^2} \right)^{1/3} r \right], \tag{3.65}$$

where $v_{max} = \frac{2a^2}{3z^{1/3}}$ and $a \simeq 0.8255 \frac{J^{1/3}}{(\rho\mu)^{1/6}}$. It should be noted that this jet velocity profile is axisymmetric, similar to a Gaussian probability distribution. It spreads such that the centerline velocity drops off as $z^{-1/3}$.

We can then also define a mass flow rate, $\dot{m}$

$$\dot{m} \simeq (36J\rho\mu z)^{1/3}, \tag{3.66}$$

such that $\dot{m}$ increases with $z^{1/3}$ as the jet entrains ambient (irrotational) fluid by dragging it along and into the jet itself.

This solution works for larger values of z away from the jellyfish. However, the solution collapses near the jet exit where it falsely implies that $\dot{m} = 0$ at $z = 0$. Thus, we conclude that the self-similar solution does not work in developing region near the jellyfish bell itself, where Re_z is small.

If we were to experimentally measure the velocity profile in the far field ($z = 5d$) and obtain the following expression for the velocity profile:

$$v_z(r) = 0.5 \operatorname{sech}^2 [422.4r] \quad \text{at} \quad z = 0.5\text{m}, \tag{3.67}$$

we would even be able to determine the thrust produced by the jellyfish in question. Let us start by drawing a control volume around the body and then along streamlines where $v_z \simeq U_\infty$, as shown in Fig. 3.15.

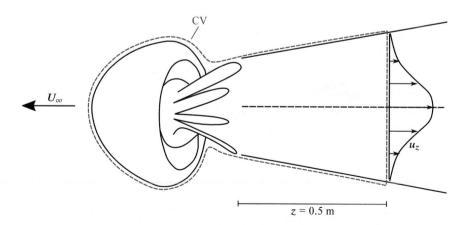

Fig. 3.15 Control volume drawn around the jellyfish and all fluid within the streamlines defined by $v_z \simeq U_\infty$

Since the thrust T is effectively the momentum flux across the control volume surface, we obtain this final expression:

$$T = \sum F_z = 2\pi \int_{-r}^{r} \rho(U_\infty)^2 r\,dr + 2\pi \int_{-r}^{r} \rho v_z^2 r\,dr, \qquad (3.68)$$

where the first term on the right-hand side drops out if the surrounding fluid is in fact quiescent (stagnant).

3.3.2 Exercise on Bounding Flight and Wakes

Consider a chickadee in bounding flight in Fig. 3.16.

Bounding flight is described by periods of short flapping bursts followed by a glide with the wings tucked back against the body. The bird will then open its wings again to slow down and perch. During the glide portion of the bird's flight (t_0), the velocity in the wake five body lengths behind the bird (where it can be assumed pressure has recovered) can be described by the following function:

$$v_z(r, z) = U_\infty - z^{-1/2} e^{\left(\frac{-U_\infty r^2}{4z\upsilon}\right)}, \qquad (3.69)$$

where U_∞ is the freestream velocity and υ is the kinematic viscosity of air. Assuming the chickadee is 10 cm long and traveling at 5 m/s, let us evaluate $v_z(r, z)$ and calculate the vorticity field at $z = 0.5$ m.

Assuming axisymmetric, planar flow and recalling that the vorticity is simply the curl of the velocity field:

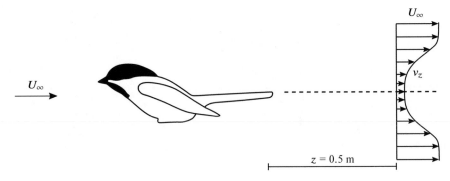

Fig. 3.16 The wake of a tiny chickadee (approximately 0.1 m long) during the bounding portion of its flight

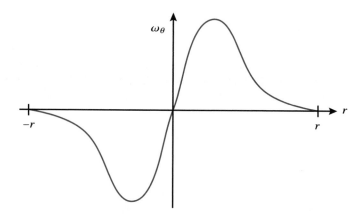

Fig. 3.17 The vorticity profile is indicative of counter-rotating vortices on each side of the chickadee

$$\boldsymbol{\omega} = \nabla \times \mathbf{v} = -\frac{\partial v_z}{\partial r}\hat{\theta}, \tag{3.70}$$

where the only non-zero component of the vorticity is that acting in the θ direction. Completing this operation produces the vorticity field, shown in Fig. 3.17:

$$\omega_\theta = \frac{5\sqrt{2}}{v}re^{-\frac{5r^2}{2v}}. \tag{3.71}$$

Now, let us consider the assumptions made in our boundary-layer approximation and apply them to a point in the wake six body lengths behind the gliding bird. What are the dominant terms in this flow?

At $z = 0.6\,\text{m}$, we can consider to be in the far field and $\delta \ll L$ is easily satisfied. The boundary-layer approximation further requires that Re$\gg$ 1:

$$\text{Re} = \frac{U_\infty L}{v} \approx 33000. \tag{3.72}$$

With each of these conditions satisfied, we can continue to make assumptions similar to those made in the previous exercise for a jet:

- $v_\theta = 0 \rightarrow$ axisymmetric flow ($\partial/\partial\theta = 0$).
- $v_r \ll v_z \rightarrow$ velocity profile dominated by body motion in the z-direction.
- $\dfrac{\partial v_z}{\partial z} \ll \dfrac{\partial v_z}{\partial r}$ and $\dfrac{\partial v_r}{\partial z} \ll \dfrac{\partial v_r}{\partial r} \rightarrow$ gradient much stronger in $\hat{r}$.
- $\partial/\partial t = 0 \rightarrow$ steady flow.

Again, following a similar procedure to the jellyfish exercise, the flow field within the wake can be described by

$$v_r \frac{\partial v_z}{\partial r} + v_z \frac{\partial v_z}{\partial z} = v \left[\frac{1}{r} \frac{\partial}{\partial r} \left(r \frac{\partial v_z}{\partial r} \right) \right], \tag{3.73}$$

where the two terms on the left-hand side are inertial, while the right-hand side accounts for viscous effects. Note that this equation is identical to the equation we derived for a jet (when the laminar boundary-layer approximation is satisfied).

3.3.3 Exercise with Unsteady Boundary Layer

One could easily argue that most larger-scale biological flows are in fact unsteady (e.g., pulsatile or oscillatory) and that steady solutions become the exception rather than the rule. As shown in Fig. 3.18, there is no doubt that often assuming steady conditions provides a nice easy way to start out with one's analysis, but if this approach cuts out key aspects of the flow, then we will ultimately miss out on important flow physics. In order to delve into this fascinating unsteady world, perhaps the best first taste lies with Stokes' second problem (see Fig. 3.19),

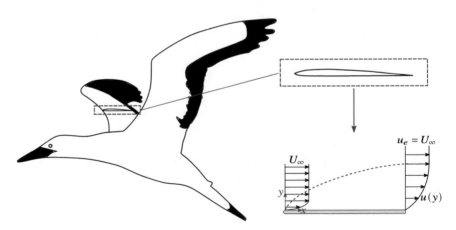

Fig. 3.18 Here the thin bird-wing cross section will be approximated as a simplified thin flat plate with a growing boundary layer, as shown in the bottom right

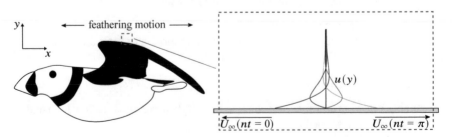

Fig. 3.19 Consider the oscillating boundary layer as per Stokes' second problem

which concerns itself with the time-varying boundary layer over an infinitely long oscillating plate (Stokes 1843). One should note that this problem concerns inertial effects and must not be confused with Stokes' flow.

We start by simplifying the Navier–Stokes equation as per the boundary-layer approximation but with careful attention to the unsteady term, among other constraints. Let us consider again the bulk two-dimensional flow where for x-momentum:

$$\frac{\partial u}{\partial t} + u \frac{\partial u}{\partial x} + v \frac{\partial u}{\partial y} = -\frac{1}{\rho} \frac{\partial p}{\partial x} + v \left(\frac{\partial u^2}{\partial x^2} + \frac{\partial u^2}{\partial y^2} \right). \tag{3.74}$$

It is important in this current problem to recognize that the boundary conditions are strongly time-dependent, and as such, the flow is unsteady, such that $St \neq 0$, and therefore

$$\frac{\partial u}{\partial t} \neq 0. \tag{3.75}$$

Furthermore, because the flat plate is infinitely long, there is no dependency (variation) with respect to x such that:

$$\frac{\partial u}{\partial x} = 0, \tag{3.76}$$

and hence

$$\frac{\partial u^2}{\partial x^2} = 0. \tag{3.77}$$

For the wall-normal direction, where $v = 0$, we obtain

$$v \frac{\partial u}{\partial y} = 0. \tag{3.78}$$

Finally, we know when examining the y-component of momentum through the lens of the boundary-layer approximation that

$$\frac{\partial p}{\partial y} = 0, \tag{3.79}$$

or simply $p \neq f(y)$, such that

$$\frac{\partial p}{\partial x} = \frac{dp}{dx}. \tag{3.80}$$

However, in the far field ($y \neq 0$), the fluid is at rest such that by definition pressure cannot vary parallel to the wall either:

$$\frac{dp}{dx} = 0. \tag{3.81}$$

In other words, pressure is constant throughout the domain in this unusual problem, and to conclude, the Navier–Stokes equation simplifies as follows:

$$\frac{\partial u}{\partial t} = v\frac{\partial u^2}{\partial y^2}. \tag{3.82}$$

This expression takes on an analogous form to the general diffusion equation, or in heat-transfer what is known as the *heat equation*:

$$\frac{\partial T}{\partial t} = D\frac{\partial T^2}{\partial y^2}, \tag{3.83}$$

where T represents temperature and D is the diffusion coefficient. In order to solve the parabolic PDE (Eq. (3.82)), we apply the *separation of variables*, where the general solution takes on the form:

$$u(y, t) = U_o e^{-ky} \cos(nt - ky). \tag{3.84}$$

U_o is the amplitude of oscillation, n the frequency of oscillation, and k must be calculated by substituting the velocity $u(y, t)$ into Eq. (3.82) and then simplifying, such that

$$k = \sqrt{\frac{n}{2v}}. \tag{3.85}$$

In a more compact form, we define

$$\eta = ky = y\sqrt{\frac{n}{2v}} \tag{3.86}$$

and the solution for the oscillating velocity profile as

$$u(y, t) = U_0 e^{-\eta} \cos(nt - \eta). \tag{3.87}$$

We observe here that $e^{-\eta}$ is a damping function responsible for diminishing the oscillations imposed by the wall. In a similar vein, $\cos(nt)$ is a harmonic oscillation, such that when combined $e^{-\eta} \cos(nt - \eta)$ represents a damped harmonic oscillation. It also should be noted that the η term in $\cos(nt - \eta)$ represents a phase lag with respect to the motion of the wall. The lag is of the form $\eta = y\sqrt{\frac{n}{2v}}$ and is affected by

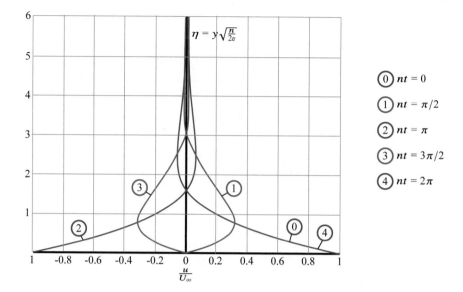

Fig. 3.20 Varying solutions to Stokes' second problem

the fluid viscosity and oscillation frequency as the distance y from the wall varies. We can examine the various velocity profiles for different phases (time) in Fig. 3.20.

3.4   Turbulent Boundary-Layer Equation

Now that we have explored the boundary-layer approximation, let us consider what happens as we continue to increase the Reynolds number and enter the realm of transitional and turbulent flows. Recall that in general when Re $\gg$ 1, we obtain the following for 2D flow:

- Continuity

$$\frac{\partial u}{\partial x} + \frac{\partial v}{\partial y} = 0, \tag{3.88}$$

- y-Momentum

$$\frac{\partial p}{\partial y} \cong 0, \tag{3.89}$$

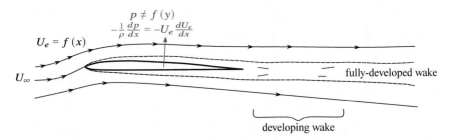

Fig. 3.21 Consider the attached flow over an airfoil (two-dimensional section of a wing). Let us imagine this is the flow over a seagull's wing under gliding (steady) conditions

- x-Momentum

$$u\frac{\partial u}{\partial x} + v\frac{\partial u}{\partial y} = -\frac{1}{\rho}\frac{dp}{dx} + v\frac{\partial^2 u}{\partial y^2}, \tag{3.90}$$

where $p \neq f(y)$, $-\frac{1}{\rho}\frac{dp}{dx} = -U_e\frac{dU_e}{dx}$, and $U_e = f(x)$. Let us look at the flow over a two-dimensional section of a wing, referred to as an airfoil. We can imagine this to be a snapshot of the flow over a seagull's wing under gliding (steady) conditions; see Fig. 3.21.

Once in the developed wake region, where there exists a momentum deficit (as shown in Fig. 3.22) and $\frac{\delta}{\partial x} \ll 1$ is satisfied, the x-component of momentum (Eq. (3.90)) reduces to

$$u\frac{\partial u}{\partial x} + v\frac{\partial u}{\partial y} \approx v\frac{\partial^2 u}{\partial y^2} \tag{3.91}$$

since $\frac{dp}{dx} \approx 0$ for minimal curvature (e.g., a flat plate).

Furthermore, if the velocity deficit on the wake centerline satisfies $u \ll U_\infty$, then the x-component of momentum further simplifies to

$$U_\infty\frac{\partial u}{\partial x} \approx v\frac{\partial^2 u}{\partial y^2}, \tag{3.92}$$

where $u(x, \pm\infty) = 0$ and $\frac{\partial u}{\partial y} = 0$ at $y = 0$.

For the planar (two-dimensional) wake we obtain an expression in the form of the linear heat-conduction equation, and by solving we find the velocity profile:

$$u(x, y) = BU_\infty x^{-\frac{1}{2}}exp\left(-\frac{U_\infty y^2}{4xv}\right), \tag{3.93}$$

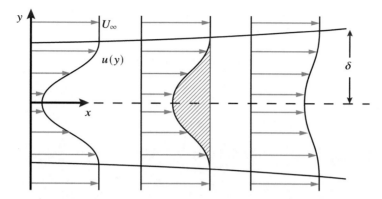

Fig. 3.22 Self-similar velocity profile in the wake of a bluff body. Blue cross-hatching represents a momentum deficit in the wake

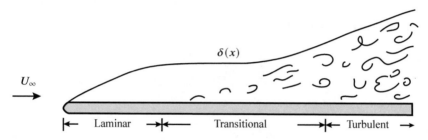

Fig. 3.23 Perturbations in the flow during transition to turbulence cause the boundary-layer thickness to grow rapidly and thus the boundary-layer approximation will break down

where $B = $ const. Note that the far-field wake has a Gaussian velocity distribution, with a wake centerline velocity recovering as $u = f(x^{-\frac{1}{2}})$.

Let us now explore how far we can go by increasing the Reynolds number before the boundary-layer approximation breaks down. Once we observe perturbations relative to the mean flow field indicative of transition to turbulence, as illustrated in Fig. 3.23, we decompose the flow into mean and fluctuating motions, referred to as the Reynolds decomposition such that:

$$u = \bar{u} + u',\, v = \bar{v} + v',\, w = \bar{w} + w' \text{ and } p = \bar{p} + p', \tag{3.94}$$

where, for instance:

$$\bar{u} = \frac{1}{t_1} \int_{t_0}^{t_0+t_1} u\, dt. \tag{3.95}$$

We then integrate over a long time interval t_1 such that the mean values become independent of time, and then by definition we observe that:

$$\bar{u'} = 0, \bar{v'} = 0, \bar{w'} = 0 \text{ and } \bar{p'} = 0. \tag{3.96}$$

Tips

We observe the following at this stage:

- Decomposition applies relative to a *steady* mean flow although, in principle, this could be extended to time-varying (unsteady) flows as well; and
- Fluctuations act about the local mean flow by effectively increasing the apparent mixing, which in turn can be imagined as a local increase in viscosity.

In a transitional or turbulent regime, the continuity equation remains unchanged:

$$\frac{\partial \bar{u}}{\partial x} + \frac{\partial \bar{v}}{\partial y} + \frac{\partial \bar{w}}{\partial z} = 0 \tag{3.97}$$

and, similarly, we observe that

$$\frac{\partial u'}{\partial x} + \frac{\partial v'}{\partial y} + \frac{\partial w'}{\partial z} = 0. \tag{3.98}$$

To address momentum, we must now rearrange the Navier–Stokes equation to obtain what is referred to as the *Reynolds-averaged* Navier–Stokes equation. Here we observe a series of new terms (on the far right-hand side) that are formed so as to account for the additional transfer of momentum associated with turbulent fluctuations:

$$\bar{u}\frac{\partial \bar{u}}{\partial x} + \bar{v}\frac{\partial \bar{u}}{\partial y} + \bar{w}\frac{\partial \bar{u}}{\partial z} = -\frac{1}{\rho}\frac{\partial \bar{p}}{\partial x} + v\left(\frac{\partial^2 \bar{u}}{\partial x^2} + \frac{\partial^2 \bar{u}}{\partial y^2} + \frac{\partial^2 \bar{u}}{\partial z^2}\right)$$

$$- \left(\frac{\partial \overline{u'^2}}{\partial x} + \frac{\partial \overline{u'v'}}{\partial y} + \frac{\partial \overline{u'w'}}{\partial z}\right)$$

$$\bar{u}\frac{\partial \bar{v}}{\partial x} + \bar{v}\frac{\partial \bar{v}}{\partial y} + \bar{w}\frac{\partial \bar{v}}{\partial z} = -\frac{1}{\rho}\frac{\partial \bar{p}}{\partial y} + v\left(\frac{\partial^2 \bar{v}}{\partial x^2} + \frac{\partial^2 \bar{v}}{\partial y^2} + \frac{\partial^2 \bar{v}}{\partial z^2}\right) \tag{3.99}$$

$$- \left(\frac{\partial \overline{u'v'}}{\partial x} + \frac{\partial \overline{v'^2}}{\partial y} + \frac{\partial \overline{v'w'}}{\partial z}\right)$$

$$\bar{u}\frac{\partial \bar{w}}{\partial x} + \bar{v}\frac{\partial \bar{w}}{\partial y} + \bar{w}\frac{\partial \bar{w}}{\partial z} = -\frac{1}{\rho}\frac{\partial \bar{p}}{\partial z} + v\left(\frac{\partial^2 \bar{w}}{\partial x^2} + \frac{\partial^2 \bar{w}}{\partial y^2} + \frac{\partial^2 \bar{w}}{\partial z^2}\right)$$
$$-\left(\frac{\partial \overline{u'w'}}{\partial x} + \frac{\partial \overline{v'w'}}{\partial y} + \frac{\partial \overline{w'^2}}{\partial z}\right).$$

These new terms on the right-hand side are referred to as Reynolds stresses and are challenging to predict or model since transitional and turbulent flows can be highly *anisotropic*, i.e. turbulence often acts in a non-uniform manner in space.

Alternatively, we can express the above Reynolds-averaged Navier–Stokes equation in a much more compact form through Einstein notation:

$$\frac{\partial \bar{u}_i}{\partial x_i} = 0$$

$$\bar{u}_j \frac{\partial \bar{u}_i}{\partial x_j} = -\frac{1}{\rho}\frac{\partial \bar{p}}{\partial x_i} + v\frac{\partial^2 \bar{u}_i}{\partial x_j \partial x_j} - \frac{\partial \overline{u'_i u'_j}}{\partial x_j}. \tag{3.100}$$

As with the boundary-layer approximation, the turbulent boundary-layer equation's character is analogous. For instance, continuity shares a similar form in two-dimensions such that

$$\frac{\partial \bar{u}}{\partial x} + \frac{\partial \bar{v}}{\partial y} = 0. \tag{3.101}$$

As the Reynolds number is systematically increased, we observe that $\bar{v} \ll U_\infty$ and $\frac{\partial}{\partial x} \ll \frac{\partial}{\partial y}$. Furthermore, the mean flow is, as before, assumed to be two-dimensional such that $\bar{w} = 0$ and $\frac{\partial}{\partial z} = 0$. Thus, in the end, we obtain

$$0 = -\frac{1}{\rho}\frac{\partial \bar{p}}{\partial y} - \frac{\partial \overline{v'^2}}{\partial y}. \tag{3.102}$$

We can then integrate this expression over the boundary-layer thickness to obtain

$$\bar{p} + \rho\overline{v'^2} = \bar{p}_w = p_e, \tag{3.103}$$

where $\bar{p}_w$ and p_e represent the pressure at the wall and in the outer region (free of turbulence), respectively.

Finally, for the x-component of momentum we obtain the following expression:

$$\bar{u}\frac{\partial \bar{u}}{\partial x} + \bar{v}\frac{\partial \bar{u}}{\partial y} = -\frac{1}{\rho}\frac{dp_e}{dx} + \frac{\partial}{\partial y}(\bar{\tau}_v + \tau_t), \tag{3.104}$$

where τ_t represents the turbulent shear stresses. More commonly we present these additional *Reynolds stresses* (tau_t) in the following form (see last term):

$$\bar{u}\frac{\partial \bar{u}}{\partial x} + \bar{v}\frac{\partial \bar{u}}{\partial y} = -\frac{1}{\rho}\frac{dp_e}{dx} + \frac{\partial}{\partial y}(\mu\frac{\partial \bar{u}}{\partial y} - \rho\overline{u'v'}). \tag{3.105}$$

Tips

With the Reynolds decomposition we now observe the following:

- $\bar{\tau}_v$ represent the viscous stresses through diffusion (as previously seen for laminar flows); and
- τ_t appears as the Reynolds stresses, representing the effect of turbulent mixing on the flow.

3.4.1   Exercise for Drag on Streamlined Bodies

Let us consider the flow over the top surface of a thin streamlined body, such as an airfoil (see Fig. 3.24), where the overall drag can be assumed to be dominated by the wall shear parallel to this surface rather than the pressure drag acting in the normal direction. As in earlier examples, gliding flight, in which the unsteady/acceleration term in the Navier–Stokes equation can be neglected (i.e., $St \approx 0$), represents not only a key form of locomotion across a broad range of disparate animal groups but also a hotly debated subject in terms of the various independent origins of flight. For instance, as discussed by Dececchi et al. (2016), did the first birds run, flap, and then glide, or did they climb trees and glide down from above?

Let us consider the following aspects of this problem, where we assume bulk two-dimensional flow (b is unit length in the spanwise direction) and (as usual) incompressible conditions:

- At what rate, $\dot{m}_2$, does mass leave the control volume?
- What is the drag force, D, imparted on the fluid by the boundary layer (equal and opposite to the drag on the airfoil)?

Begin by drawing a control volume (CV), whose four sides are labeled 1–4 in Fig. 3.24, extending back from the leading edge of the airfoil. We then apply conservation of mass to the control volume so as to determine the rate that mass exits as a result of the boundary layer:

$$\left(\frac{dm}{dt}\right)_{sys} = 0 = \frac{d}{dt}\left(\int_{CV}\rho d\Psi\right) + \int_{CS}\rho(\mathbf{v}\cdot\mathbf{n})dA, \tag{3.106}$$

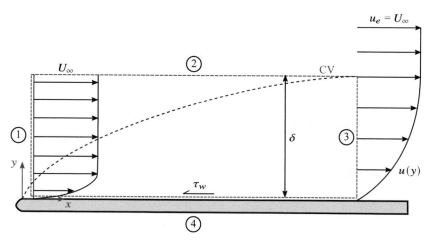

Fig. 3.24 Approximation of drag on a thin, streamlined body with control volume. For this exercise we assume a two-dimensional flow representative of an inboard section of a wing

where the unsteady term vanishes in a steady flow and

$$\int_{CS} \rho(\mathbf{v} \cdot \mathbf{n}) dA = 0, \tag{3.107}$$

where $dA = bdy$. CS represents the *control surface* and, hence, the sum of all fluxes ($\dot{m}$) across the control surfaces (1–4) will sum to zero:

$$\dot{m}_{out} - \dot{m}_{in} = 0. \tag{3.108}$$

Assuming a mean two-dimensional flow, i.e., $v_z = 0$, and mass only entering through surface 2, we can write

$$\cancel{\dot{m}_4} + \dot{m}_2 + \rho b \int_0^\delta u(y) dy - \rho b \int_0^\delta U_\infty dy = 0, \tag{3.109}$$

where $\dot{m}_2$ is our unknown and the airfoil is non-porous ($\dot{m}_4 = 0$). We can then rearrange to obtain

$$\dot{m}_2 = \rho b \int_0^\delta (u(y) - U_\infty) \, dy = \rho b U_\infty \int_0^\delta \left(1 - \frac{u(y)}{U_\infty}\right) dy, \tag{3.110}$$

such that

$$\frac{\dot{m}_2}{\rho b U_\infty} = \int_0^s \left(1 - \frac{u(y)}{U_\infty}\right) dy = \delta^*. \tag{3.111}$$

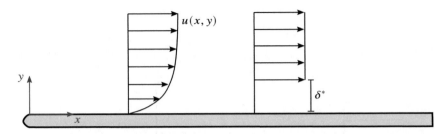

Fig. 3.25 The displacement thickness δ^* is a measure of the effective thickness of the streamlined body. The less drag on the surface, say through semi-slip, the less the flow will have to displace laterally away from the body

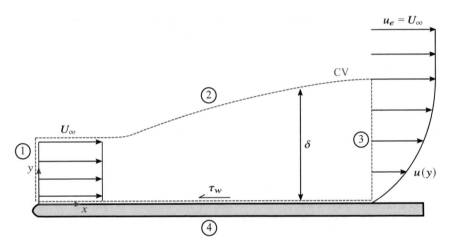

Fig. 3.26 By selecting instead the upper bounding streamline as a control surface (through which no x-momentum passes) we can calculate the drag on our airfoil more easily

δ^* is commonly referred to as the *displacement thickness*, which can be thought of as representing the effective thickness of the streamlined body, as shown in Fig. 3.25.

To calculate the drag force (D), we now apply conservation of momentum to a new control volume such that no x-momentum escapes through the upper control surface, as shown in Fig. 3.26.

Through the x-momentum balance we obtain the following, again striking out the unsteady term for this steady (gliding) condition:

$$\frac{d}{dt}(m\mathbf{v})_{sys} = \sum \mathbf{F}_x = \frac{d}{dt}\left(\int_{CV} \mathbf{v}\rho d\mathbf{v}\right) + \int_{CS} \mathbf{v}\rho(\mathbf{v} \cdot \mathbf{n})dA. \qquad (3.112)$$

Since in this abstracted problem we assume that there is no pressure force acting in the x-direction, any momentum flux will be equal to the drag force:

$$-D = b \int_0^\delta u(\rho u)dy - b \int_0^h U_\infty(\rho U_\infty)dy = 0, \qquad (3.113)$$

and hence

$$\frac{D}{\rho b} = U_\infty^2 h - \int_0^\delta u^2(y)dy, \qquad (3.114)$$

where h represents the unknown inlet height of our new control volume, as shown in Fig. 3.26. In order to determine h we must perform another mass balance:

$$\rho \int_{CS} \mathbf{u} \cdot \hat{n} dA = 0 \quad \rightarrow \quad \rho b \int_0^\delta u(y)dy - \rho b \int_0^h U_\infty dy = 0, \qquad (3.115)$$

such that we obtain

$$U_\infty h = \int_0^\delta u(y)dy. \qquad (3.116)$$

We then substitute the above expression back into Eq. (3.114) to determine the total drag on this upper (suction) surface:

$$\frac{D}{\rho b} = U_\infty \int_0^\delta u(y)dy - \int_0^\delta u^2(y)dy$$

$$= \int_0^\delta u(U_\infty - u)dy \qquad (3.117)$$

$$\frac{D}{\rho b U_\infty^2} = \int_0^\delta \frac{u}{U_\infty}\left(1 - \frac{u}{U_\infty}\right)dy = \theta,$$

where θ is the *momentum thickness* and represents the amount of momentum transferred into friction at the wall.

Finally, the sectional (two-dimensional) drag coefficient, C_D, is determined simply by normalizing using the airfoil chord length c:

$$C_D = \frac{D}{\frac{1}{2}\rho U_\infty^2 bc}. \qquad (3.118)$$

When applying the known Blasius solution for laminar boundary layers (Blasius 1908), we obtain an expression based on Reynolds number as determined by the plate (or chord) length:

$$C_D = \frac{1.328}{\text{Re}_c{}^{1/2}}. \tag{3.119}$$

One interesting observation is that the drag coefficient is then observed to decrease with increasing freestream velocity (Re_c) for laminar flow in these simplified abstractions of steady lifting surfaces pertinent to gliding.

Now, what about drag on this suction surface when instead a turbulent boundary layer sets itself up at much larger scales (higher Re_c)? The procedure to determine the drag becomes much more complicated to be performed analytically, but luckily semi-empirical relations are available (Schlichting & Gersten 2016), for example:

$$C_D = \frac{0.455}{(\log_{10}\text{Re}_c)^{2.58}}. \tag{3.120}$$

As per Eqs. (3.119) and (3.120), we observe that C_D grows inversely proportional to Re_c (as we vary U_∞ in both laminar and turbulent regimes).

The process of transition from laminar to turbulent conditions, which even for this incredibly simple case with zero pressure gradient in the streamwise direction, is highly complex and actively researched. In particular, for biological swimming and flying, specific surface properties, such as passive or active flexibility and patterned or random roughness, can be used to either trigger or dampen perturbations that lead to turbulence downstream. If for the moment we assume the most basic situation, where perhaps turbulent flow transitions at the leading edge at approximately $\text{Re}_{tr} = 500{,}000$, we obtain the following crude delineation between two scenarios. The first scenario is such that we observe a complete laminar boundary layer along the airfoil surface versus a second scenario where we instead observe a complete turbulent boundary layer from leading to trailing edges:

$$\begin{aligned} \text{laminar } C_D &= \frac{1.328}{\sqrt{500000}} = 0.0019 \\[2mm] \text{turbulent } C_D &= \frac{0.031}{(500000)^{1/7}} = 0.0048. \end{aligned} \tag{3.121}$$

This result shows us empirically, based on measurements from a smooth surface, that the drag coefficient jumps for the turbulent regime. Furthermore, the dimensional drag (D), which the animal ultimately needs to overcome, grows considerably with increasing Reynolds number, particularly in the turbulent regime. It should be noted that these results are analogous to the trends in the friction factor observed for pipe flow, as tabulated in the Moody diagram (Moody 1944). We will address these similar observations in the next chapter on internal flows.

References

Blasius, H. (1908). Grenzschichten in Flüssigkeiten mit kleiner Reibung. *Zeitschrift für angewandte Mathematik und Physik, 56*, 1–37.

Buckingham, E. (1915). The principle of similitude. *Nature, 96*(2406), 396–397.

Dececchi, T. A., Larsson, H. C., & Habib, M. B. (2016). The wings before the bird: An evaluation of flapping-based locomotory hypotheses in bird antecedents. *PeerJ, 4*, e2159.

Moody, L. (1944). Friction factors for pipe flow. *Transactions of the American Society Mechanical Engineers, 66*(8), 671–678.

Schlichting, H., & Gersten, K. (2016). *Boundary-layer theory*. Springer Berlin Heidelberg.

Stokes, G. G. (1842). On the steady motion of incompressible fluids. *Transactions of the Cambridge Philosophical Society, 7*, 439–453.

Stokes, G. G. (1843). On some cases of fluid motion. *Transactions of the Cambridge Philosophical Society, 8*, 105–137.

Chapter 4
Internal Flows

Now that we have explored the full range of scales in Chap. 3 and considered their respective mathematical treatment, let us apply this knowledge and delve into the world of internal flows, where more often than not the flow is pulsatile and walls— sometimes even flexible walls—bound our problem. When getting into these topics, we often focus on human-centric perspectives (e.g., cardiovascular and pulmonary flows), along with some exploration of non-Newtonian properties (assumptions) pertinent to our blood and other complex fluids in our bodies. While the literature is dominated by the biomedical community, and naturally focused on human-based problems, we should not lose sight of our origins (i.e., the likely *suboptimal* solutions through phylogenetic inertia). For instance, by contrasting human hearts with other mammalian hearts, and then possibly those found in invertebrates— e.g., octopuses have three!—we might end up learning quite a bit more than when we study each in arbitrary isolation. To that end, we start out by exploring the properties of non-linear fluids before delving into increasingly complex problems containing steady, pulsatile, and even coupled fluid–structure behavior pertinent to cardiovascular and pulmonary flows.

4.1 Couette Flow

Let us begin by exploring some of the basic properties of strange fluids like blood, which after all is arguably the most critical fluid to animal life. To consider the very basic viscous properties of a dense suspension (almost half of blood by volume is cells), we must properly define shear versus strain-rate relationships in what are known as Newtonian and non-Newtonian (analogous), single-phase fluids. Couette (1890) performed simple yet invaluable experiments exploring the flow between fixed and rotating concentric cylinders, as shown in Fig. 4.1. It should be noted that modern day rheometry, whether studying steady or even time-varying properties of

© The Author(s), under exclusive license to Springer Nature Switzerland AG 2022
D. E. Rival, *Biological and Bio-Inspired Fluid Dynamics*,
https://doi.org/10.1007/978-3-030-90271-1_4

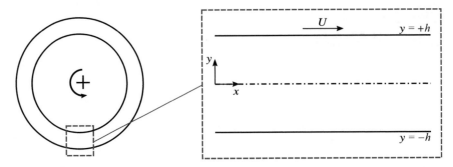

Fig. 4.1 Flow between concentric cylinders, where the inner cylinder is rotating and the outer cylinder is stationary

deforming materials, still very much relies on the basic principles of Couette flow described herein.

Consider the motion of a fluid between two concentric cylinders (one rotating and one fixed) that are separated by a short distance $2h$, as depicted in Fig. 4.1. When we zoom in sufficiently, the problem becomes effectively planar and therefore from (2D) continuity:

$$\frac{\partial u}{\partial x} + \frac{\partial v}{\partial y} = 0, \tag{4.1}$$

we observe that $u = u(y)$, given $v = 0$. In that vein, let us return to the x-momentum equation:

$$u\frac{\partial u}{\partial x} + v\frac{\partial u}{\partial y} = -\frac{1}{\rho}\frac{\partial p}{\partial x} + v\left(\frac{\partial^2 u}{\partial x^2} + \frac{\partial^2 u}{\partial y^2}\right), \tag{4.2}$$

where ultimately all terms must vanish such that:

$$\frac{\partial^2 u}{\partial y^2} = 0. \tag{4.3}$$

We can now integrate twice and obtain the following:

$$u = C_1 y + C_2, \tag{4.4}$$

where both constants are found by applying the no-slip condition:

$$u(-h) = 0 \text{ and } u(+h) = U, \tag{4.5}$$

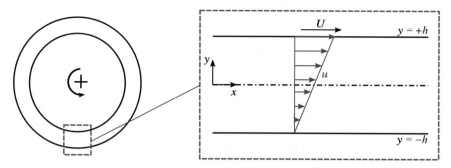

Fig. 4.2 Linear velocity distribution in Couette flow

on both cylinder walls such that

$$C_1 = \frac{U}{2h} \text{ and } C_2 = \frac{U}{2}. \tag{4.6}$$

We thus obtain the following velocity distribution (as shown in Fig. 4.2):

$$u(y) = \frac{U}{2}\left(1 + \frac{y}{h}\right), \tag{4.7}$$

where we observe that

$$\frac{du}{dy} = \frac{U}{2h} = \text{const.} \tag{4.8}$$

For the current case, the strain rate (as defined in Sect. 2.4) simplifies to

$$\epsilon_{xy} = \frac{1}{2}\left(\frac{\partial u}{\partial y} + \frac{\partial v}{\partial x}\right) = \frac{1}{2}\frac{du}{dy}. \tag{4.9}$$

Through empiricism, it has been found that, for most common fluids, the applied shear is a unique function of strain rate where

$$\tau_{xy} = f(\epsilon_{xy}). \tag{4.10}$$

For simple (Newtonian) fluids, this relationship is linear such that:

$$\tau_{xy} = \mu\frac{U}{2h} = \mu\frac{du}{dy} = 2\mu\epsilon_{xy}. \tag{4.11}$$

But how about more complex biological fluids such as blood, which are often generalized as *non-Newtonian fluids*? We will explore these cases next.

4.2   Non-Newtonian and Viscoelastic Fluids

We observe that blood and other complex fluids present curious properties, as shown in Fig. 4.3. Furthermore, some of these non-Newtonian fluids may even present *hysteretic* (time-dependent) properties, as can be observed in Fig. 4.4.

Often a power-law approximation is used to describe the time-averaged bulk properties of non-Newtonian fluids and can be found in different forms such as

$$\tau_{ij} \approx 2k\epsilon_{ij}^{n-1} \quad \text{or} \quad \tau_{ij} \approx -m\epsilon_{ij}^{n}, \tag{4.12}$$

where k, m, and n are material properties measured using rheometry.

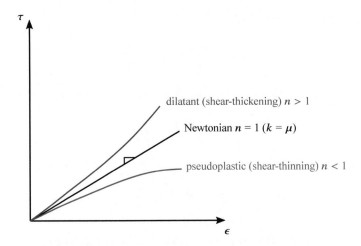

Fig. 4.3 Stress–strain curves for some Newtonian and non-Newtonian materials

Fig. 4.4 Hysteresis can be observed in some cases as well

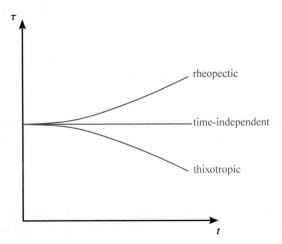

For the case of laminar flow in a pipe with a circular cross-section:

$$v_z = f(r),$$ (4.13)

such that $\epsilon_{rz} = dv_z/dr$ and Eq. (4.12) becomes

$$\tau_{rz} = -m \left(\frac{dv_z}{dr} \right)^n.$$ (4.14)

From considering a momentum balance in steady pipe flow (as will be shown in Sect. 4.4), we can relate the shear stress to a pressure drop Δp over some distance L:

$$\tau_{rz} = \frac{\Delta p}{2L} r,$$ (4.15)

such that:

$$-m \left(\frac{dv_z}{dr} \right)^n = \frac{\Delta p}{2L} r.$$ (4.16)

We can then manipulate, integrate, and set the no-slip boundary condition at $r = R$ in order to obtain

$$v_z = \left(\frac{\Delta p R}{2mL} \right)^{\frac{1}{n}} \frac{R}{(1/n) + 1} \left[1 - \left(\frac{r}{R} \right)^{(1/n)+1} \right].$$ (4.17)

When $m = \mu$ and $n = 1$ (Newtonian condition), the pipe-flow velocity profile reduces to the Hagen–Poiseuille solution (Hagen 1839; Poiseuille 1840, 1841)—as will be derived in Sect. 4.4—where

$$v_z = \frac{\Delta p R^2}{4\mu L} \left[1 - \left(\frac{r}{R} \right)^2 \right].$$ (4.18)

For example, as shown in Fig. 4.5, the laminar velocity profile in a pipe with circular cross-section will vary strongly as a function of the fluid's rheological properties. For instance, a shear-thickening fluid will present an even pointier profile than for the Newtonian case, while the profile in a shear-thinning fluid will be flatter.

Fig. 4.5 Laminar velocity profiles in a pipe with varying rheological properties

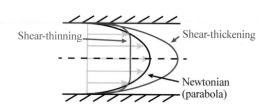

Let us now consider two-phase flows such as blood more carefully, by fully accounting for the second (solid) phase rather than using a simpler bulk approximation. Naturally, it is always more convenient to replace multi-phase flows, such as solid suspensions, with an equivalent one-phase system with an effective viscosity μ_{eff}. Such a bulk approximation can be satisfactory for some steady flows, but for many unsteady biological flows, two-phase systems must be considered as true *viscoelastic* materials.

Einstein (1906, 1911) developed an expression for a *dilute* suspension of rigid spheres:

$$\frac{\mu_{\text{eff}}}{\mu_o} = 1 + \frac{5}{2}\Phi, \tag{4.19}$$

where μ_o is the viscosity of the suspending medium (liquid) and Φ represents the volume fraction of spheres. Note that the term dilute suspension, in contrast to dense suspensions, assumes that there is no particle–particle interactions ($0.05 \lesssim \Phi$). Several semi-empirical models based on the above also exist for elongated or flexible particles.

For concentrated or dense suspensions, where $\Phi \gtrsim 0.05$, particle–particle interactions become appreciable. One semi-empirical model for rigid spheres is the Mooney equation (Mooney 1951):

$$\frac{\mu_{\text{eff}}}{\mu_o} = exp\left(\frac{\frac{5}{2}\Phi}{1 - (\Phi/\Phi_o)}\right), \tag{4.20}$$

where Φ_o represents an empirical constant based on packing density ($0.52 < \Phi_o < 0.74$).

Tips

Note, when considering dense suspensions, the limit for the packing of solid spherical particles is as follows:

- $\Phi_o = 0.52$ cubic packing; and
- $\Phi_o = 0.74$ closest packing.

Now, when considering unsteady problems, e.g., pulsatile blood flow in smaller arteries (as will be discussed in Sect. 4.5), we must now include the time derivative to account for hysteresis. In order to consider such problems, we must combine the Newtonian viscosity properties of the carrier fluid (e.g., blood plasma):

$$\tau_{ij} = \mu\epsilon_{ij}, \tag{4.21}$$

with the Hookean elasticity of the suspended particles (e.g., red blood cells)

$$\tau_{ij} = G\gamma_{ij}, \tag{4.22}$$

where G and γ_{ij} are the elastic modulus and the material strain, respectively. We should recognize that a Hookean (elastic) material in isolation has perfect memory, i.e., hysteresis is negligible.

The simplest linear viscoelastic model, developed by Maxwell (1867), is as follows:

$$\tau_{ij} + \lambda_1 \frac{\partial}{\partial t}\tau_{ij} = \eta_0\epsilon_{ij}, \tag{4.23}$$

where λ_1 is the relaxation time constant and η_0 is the zero shear rate viscosity. We can go one step further and obtain Jeffreys' model (Jeffreys 1931):

$$\tau_{ij} + \lambda_1 \frac{\partial}{\partial t}\tau_{ij} = \eta_0\left(\epsilon_{ij} + \lambda_2 \frac{\partial}{\partial t}\epsilon_{ij}\right), \tag{4.24}$$

where λ_2 is the retardation time constant and is used to account for a time-varying strain field as well.

4.2.1 Exercise with a Rheometer

Let us consider a *synovial* fluid, which helps to ensure lubrication in our joints. In this exercise, we will look to generalize its unknown properties using a power-law assumption. Experimental characterization will be performed by filling the concentric space between the two spinning cylinders of a rheometer, as shown in Fig. 4.6.

This rheometer functions by rotating the outer cylinder at an angular velocity Ω causing the biofluid of unknown properties to shear. By measuring the torque T_z required to maintain the rate of angular rotation, the viscous characteristics of the fluid can be determined. To begin, let us consider the ratio (κ) between inner and outer radii, as shown in Fig. 4.6. For this particular rheometer, there is a significant gap between the two cylinders, and therefore, we will have to use cylindrical coordinates. Note, if $\kappa \sim 1$, then the curvature is negligible and the problem could be treated as a set of parallel plates (Couette flow).

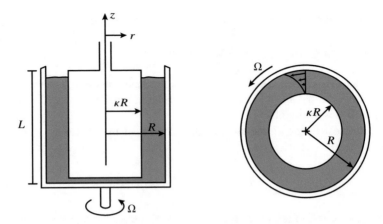

Fig. 4.6 A concentric-cylinder rheometer is used to investigate the properties of a fluid (blue) by measuring the torque required to maintain the constant angular velocity Ω of the outer cylinder

We start by defining the velocity profile for the steady, fully developed flow within the annular region between the spinning cylinders:

$$
\begin{aligned}
v_r &= 0, \\
v_\theta &= v_\theta(r), \\
v_z &= 0, \text{ and} \\
p &= p(r, z).
\end{aligned}
\tag{4.25}
$$

When considering (incompressible) continuity in cylindrical coordinates for our current problem, we observe the following simplifications:

$$
\nabla \cdot \mathbf{v} = \frac{1}{r}\frac{\partial}{\partial r}(rv_r) + \frac{1}{r}\frac{\partial v_\theta}{\partial \theta} + \frac{\partial v_z}{\partial z} = 0.
\tag{4.26}
$$

With regard to the Navier–Stokes equation, we include the stress tensor τ_{ij} since we can no longer assume a simple linear relationship between stress and strain rate. When considering the θ-component of momentum, we obtain

$$
\rho\left(\frac{\partial v_\theta}{\partial t} + v_r\frac{\partial v_\theta}{\partial r} + \frac{v_\theta}{r}\frac{\partial v_\theta}{\partial \theta} + v_z\frac{\partial v_\theta}{\partial z} + \frac{v_r v_\theta}{r}\right)
$$

$$
= -\frac{1}{r}\frac{\partial p}{\partial \theta} - \left[\frac{1}{r^2}\frac{\partial}{\partial r}(r^2\tau_{r\theta}) + \frac{1}{r}\frac{\partial}{\partial \theta}\tau_{\theta\theta} + \frac{\partial}{\partial z}\tau_{z\theta} + \frac{\tau_{\theta r} - \tau_{r\theta}}{r}\right] + \rho g_\theta,
\tag{4.27}
$$

where

$$\frac{\partial v_\theta}{\partial t} = 0 \tag{4.28}$$

due to steady conditions. From continuity (Eq. (4.26)), we know that

$$\frac{\partial v_\theta}{\partial \theta} = 0. \tag{4.29}$$

It is also observed that

$$\frac{\partial p}{\partial \theta} = 0, \tag{4.30}$$

such that $p = p(r, z)$. Finally, through symmetry the following term can be neglected as well:

$$\frac{\tau_{\theta r} - \tau_{r\theta}}{r} = 0. \tag{4.31}$$

Furthermore,

$$\tau_{\theta\theta} = -\eta \left[2 \left(\frac{1}{r} \frac{\partial \cancel{v_\theta}}{\partial \theta} + \frac{\cancel{v_r}}{r} \right) \right] + \left(\frac{2}{3}\eta - \kappa \right) (\cancel{\nabla \cdot \mathbf{v}}) = 0$$

$$\tau_{\theta z} = -\eta \left[\frac{1}{r} \frac{\partial \cancel{v_z}}{\partial \theta} + \frac{\partial \cancel{v_\theta}}{\partial z} \right] = 0, \tag{4.32}$$

where η is our non-Newtonian viscosity relating stress and strain rates.

Thus, returning to Eq. (4.27), we observe the following:

$$-\frac{1}{r^2} \frac{d}{dr} (r^2 \tau_{r\theta}) = 0. \tag{4.33}$$

Let us consider now a power-law representation for our synovial fluid:

$$\tau_{r\theta} = -\eta \epsilon_{r\theta}, \tag{4.34}$$

where

$$\eta = m \epsilon_{r\theta}^{n-1}, \tag{4.35}$$

and

$$\epsilon_{r\theta} = r \frac{\partial}{\partial r} \left(\frac{v_\theta}{r} \right) + \frac{1}{r} \frac{\partial \cancel{v_r}}{\partial \theta}. \tag{4.36}$$

Therefore, we obtain the following:

$$\tau_{r\theta} = -m\epsilon_{r\theta}^n$$
$$= -m\left(r\frac{d}{dr}\left(\frac{v_\theta}{r}\right)\right)^n. \tag{4.37}$$

When we substitute into Eq. (4.33), this results in

$$-\frac{1}{r^2}\frac{d}{dr}\left[r^2\left(-m\left(r\frac{d}{dr}\left(\frac{v_\theta}{r}\right)\right)^n\right)\right] = 0, \tag{4.38}$$

where m and n are the coefficients describing the tested fluid.
When we now integrate

$$r^2 m\left(r\frac{d}{dr}\left(\frac{v_\theta}{r}\right)\right)^n = C \tag{4.39}$$

and set

$$C_1 = C/m, \tag{4.40}$$

we obtain

$$\frac{d}{dr}\left(\frac{v_\theta}{r}\right) = \frac{1}{r}\left(\frac{C_1}{r^2}\right)^{1/n}. \tag{4.41}$$

If we then integrate yet again, we obtain

$$\frac{v_\theta}{r} = C_1^{1/n}\int r^{-2/n-1}dr = -\frac{nC_1^{1/n}}{2r^{2/n}} + C_2. \tag{4.42}$$

Applying the following boundary conditions:

1. With no-slip at $r = \kappa R$, we have $v_\theta = 0$ such that:

$$\frac{v_\theta}{\kappa R} = 0 = -\frac{C_1^{1/n}n}{2(\kappa R)^{2/n}} + C_2, \tag{4.43}$$

and therefore:

$$C_2 = \frac{nC_1^{1/n}}{2(\kappa R)^{2/n}}. \tag{4.44}$$

2. Whereas with no-slip at $r = R$, we have $v_\theta = R\Omega$ and Eq. (4.42) can be rewritten with Eq. (4.44):

$$\frac{v_\theta}{R} = \frac{R\Omega}{R} = -\frac{nC_1^{1/n}}{2(R)^{2/n}} + \frac{nC_1^{1/n}}{2(\kappa R)^{2/n}} = \frac{nC_1^{1/n}}{2}\left(\frac{1}{(\kappa R)^{2/n}} - \frac{1}{R^{2/n}}\right),$$

(4.45)

which, when rearranged:

$$\frac{nC_1^{1/n}}{2} = \Omega(\kappa R)^{2/n}\left(1 - \kappa^{2/n}\right)^{-1}.$$

(4.46)

Combining Eqs. (4.42) and (4.44), we obtain

$$\frac{v_\theta}{r} = \frac{nC_1^{1/n}}{2}\left(\frac{1}{(\kappa R)^{2/n}} - \frac{1}{r^{2/n}}\right)$$

(4.47)

and finally, with Eq. (4.46):

$$\frac{v_\theta}{r} = \Omega(\kappa R)^{2/n} \cdot \frac{\left(\frac{1}{(\kappa R)^{2/n}} - \frac{1}{r^{2/n}}\right)}{(1 - \kappa^{2/n})} = \Omega \cdot \frac{1 - (\frac{\kappa R}{r})^{2/n}}{1 - \kappa^{2/n}}.$$

(4.48)

Thus, the velocity profile within the annular region of the rheometer is represented by

$$\frac{v_\theta}{r\Omega} = \frac{1 - (\frac{\kappa R}{r})^{2/n}}{1 - \kappa^{2/n}},$$

(4.49)

as shown in Fig. 4.7. Note that Eq. (4.49) reduces to the Newtonian solution when $n = 1$, whilst $n < 1$ and $n > 1$ represent shear-thinning and shear-thickening fluids, respectively.

Fig. 4.7 Velocity profiles within the concentric cylinders of a rheometer as a function of fluid properties, where $n < 1$, $n = 1$, and $n > 1$ represent shear-thinning, Newtonian, and shear-thickening fluids, respectively

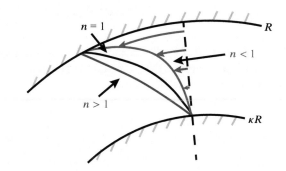

We now write an expression relating the torque, the angular rotation rate, and the material properties (m,n) of the synovial fluid. The torque, T_z, on the outer cylinder required to maintain motion is as follows:

$$T = r \times F, \tag{4.50}$$

where the force F is simply:

$$F = \tau A, \tag{4.51}$$

where A represents the surface of the cylinder wall. Thus,

$$T_z = R \cdot (-\tau_{r\theta}|_{r=R} \cdot 2\pi RL), \tag{4.52}$$

where $\tau_{r\theta}|_{r=R}$ is the shear at the wall and $2\pi RL$ is the surface area of the outer cylinder. Therefore,

$$T_z = m \left(r \frac{d}{dr} \left(\frac{v_\theta}{r} \right) \right)^n \bigg|_{r=R} \cdot 2\pi R^2 L, \tag{4.53}$$

and if we substitute in our velocity profile from Eq. (4.49):

$$
\begin{aligned}
T_z &= m \left(r \frac{d}{dr} \left(\Omega \cdot \frac{1 - (\frac{\kappa R}{r})^{2/n}}{1 - \kappa^{2/n}} \right) \right)^n \bigg|_{r=R} \cdot 2\pi R^2 L \\
&= m \left(r \left[\frac{\Omega}{1 - \kappa^{2/n}} \cdot \frac{2(\kappa R)^{2/n}}{nr^{2/n+1}} \right] \right)^n \bigg|_{r=R} \cdot 2\pi R^2 L \\
&= m \left(\frac{2\Omega (\kappa R)^{2/n}}{n(1 - \kappa^{2/n}) R^{2/n}} \right)^n \cdot 2\pi R^2 L,
\end{aligned}
\tag{4.54}
$$

we obtain finally

$$T_z = 2\pi m \Omega^n (\kappa R)^2 L \left(\frac{2/n}{1 - \kappa^{2/n}} \right)^n. \tag{4.55}$$

With this expression, we take the measured T_z data, as a function of angular velocity Ω and then determine the power-law parameters m and n for which we can describe our specific synovial fluid in question.

4.2.2 Modeling Blood: The Casson Model

Blood is a highly complex fluid. It is a two-phase non-Newtonian fluid and approximately 45–55% of its volume is occupied by suspended red blood cells. The Casson model (1959)—one of many blood flow models—attempts to describe the shear-thinning behavior exhibited by blood and takes the following form:

$$\tau(\epsilon) = \tau_0 + k^2\epsilon + 2k\sqrt{\tau_0}\sqrt{\epsilon}, \tag{4.56}$$

where τ is the shear stress, τ_0 is the yield stress, k is a material constant, and ϵ is the strain experienced by the fluid. What is the effect of each of these strain-dependent terms?

We can assume that the strain on the fluid is proportional to the Reynolds number of the flow (i.e., velocity gradients). At low Re, ϵ in the second term will be smaller than $\sqrt{\epsilon}$ in the third term. However, as Re is increased, $\epsilon \gg \sqrt{\epsilon}$ and the non-linear becomes negligible:

$$\tau = \tau_0 + k^2\epsilon, \tag{4.57}$$

where τ scales linearly with ϵ (at high Re) assuming $k = const$.

In conclusion, the Casson model is best for small vessels where flow velocities are relatively small and the various strain terms are of similar magnitude. Larger vessels, like the aorta, are perhaps more accurately represented by Newtonian models, but unsteady effects must be considered, as will be discussed in Sect. 4.5.

4.3 Steady Duct Flow

So far, we have hinted at the complexities of flows in our cardiovascular system in the form of pulsatility, turbulence, vortex rings, and valves and bifurcations. Before we attempt to solve any problems involving so many variables, we must begin with steady, laminar duct flow. Blood flow in the aorta may be anything but steady and laminar, but recall the fractal breakdown of the cardiovascular system and the conservation of mass at each level. Large arteries branch into smaller and smaller vessels, increasing the resistance to flow, which dampens oscillations and decreases velocity such that the steady and laminar assumptions are reasonable for flow in small vessels (e.g., arterioles).

Flow through a duct of arbitrary, constant cross-sectional area (e.g., the partially blocked vessel in Fig. 4.8) is governed by the incompressible continuity equation:

$$\nabla \cdot \mathbf{v} = 0, \tag{4.58}$$

Fig. 4.8 Pressure-driven flow through an occluded blood vessel section with constant cross-sectional area A. The occlusion is shaded gray

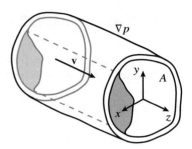

and the Navier–Stokes equation:

$$\frac{\partial \mathbf{v}}{\partial t} + (\mathbf{v} \cdot \nabla)\mathbf{v} = -\frac{1}{\rho}\nabla p + v\nabla^2 \mathbf{v} + \mathbf{F}_{body}. \tag{4.59}$$

This system of equations has no general solution, but there are some exact solutions, including (linear) Poiseuille flow.

Poiseuille was a French physician and physiologist interested in hemodynamics. In 1838, he experimentally derived a relationship—aptly named Poiseuille's law—between the flow and the pressure gradient in a long, constant cross-section tube. The relationship assumes the flow is steady, fully developed, and laminar.

> **Tips**
> *Fully developed* implies the following:
>
> - Flow is purely axial where $u = 0$, $v = 0$, $w = w(x, y)$; and
> - The velocity profile is invariable along the axis of the duct (z-direction) such that $\partial w / \partial z = 0$.
>
> A *steady* flow has no temporal variation: $\partial \mathbf{v}/\partial t = 0$.

With these assumptions, the governing equations describing the flow in Fig. 4.8 can be simplified such that continuity reduces to

$$\nabla \cdot \mathbf{v} = \frac{\partial u}{\partial x} + \frac{\partial v}{\partial y} + \frac{\partial w}{\partial z} = 0, \tag{4.60}$$

as identically satisfied by the fully developed condition.

Almost all of the terms in the y- and z-momentum equations are eliminated such that:

$$\left. \begin{array}{lll} \text{x-momentum} & \rightarrow & \dfrac{\partial p}{\partial x} = 0 \\[2mm] \text{y-momentum} & \rightarrow & \dfrac{\partial p}{\partial y} = 0 \end{array} \right\} \, p = p(z), \tag{4.61}$$

and the pressure term is a function of z only. Note that the body forces are conservative and negligible. Similarly, the z-momentum reduces to the following:

$$\cancel{\frac{\partial u}{\partial t}} + u\cancel{\frac{\partial w}{\partial x}} + v\cancel{\frac{\partial w}{\partial y}} + w\cancel{\frac{\partial w}{\partial z}} = -\frac{1}{\rho}\frac{dp}{dz} + v\left(\frac{\partial^2 w}{\partial x^2} + \frac{\partial^2 w}{\partial y^2} + \cancel{\frac{\partial^2 w}{\partial z^2}}\right)$$

$$\frac{\partial^2 w}{\partial x^2} + \frac{\partial^2 w}{\partial y^2} = \frac{1}{\mu}\frac{dp}{dz}.$$

(4.62)

Equation (4.62) is the expression for Poiseuille flow and is subject only to the no-slip condition ($w_{wall} = 0$). This result is linear and dp/dz is constant and negative. Physically, the pressure gradient must be negative along the flow direction to overcome wall shear stress.

4.4 Hagen–Poiseuille Flow

Now that we have derived the basic expression for fully developed duct flow, we can apply the same conditions to pipes with round cross-sections, which are reasonable first approximations to, for instance, the flow in our blood vessels. This first steady approximation is referred to as a Hagen–Poiseuille flow. Let us consider a section of round pipe (radius R), as shown in Fig. 4.9, and change to cylindrical coordinates where the cross-section can now be represented by r, θ instead of y, z.

As before, flow is fully developed and moving only in the axial direction where:

$$v_r = 0, v_\theta = 0.$$

(4.63)

When examining the continuity equation,

$$\frac{1}{r}\frac{d}{dr}(rv_r) + \frac{1}{r}\frac{\partial v_\theta}{\partial \theta} + \frac{\partial v_z}{\partial z} = 0,$$

(4.64)

Fig. 4.9 Description of a circular-pipe-flow geometry

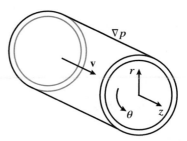

the first two terms vanish such that:

$$\frac{\partial v_z}{\partial z} = 0. \tag{4.65}$$

From the r-component of momentum,

$$\frac{\partial v_r}{\partial t} + (\mathbf{v} \cdot \nabla)v_r - \frac{1}{r}v_\theta^2 = -\frac{1}{\rho}\frac{\partial p}{\partial r} + v\left[\nabla^2 v_r - \frac{v_r}{r^2} - \frac{2}{r^2}\frac{\partial v_\theta}{\partial \theta}\right], \tag{4.66}$$

we obtain a straightforward result that

$$\frac{\partial p}{\partial r} = 0. \tag{4.67}$$

Similarly, when examining the θ-component of momentum:

$$\frac{\partial v_\theta}{\partial t} + (\mathbf{v} \cdot \nabla)v_\theta + \frac{1}{r}v_r v_\theta = -\frac{1}{\rho r}\frac{\partial p}{\partial \theta} + v\left[\nabla^2 v_\theta - \frac{v_\theta}{r^2} + \frac{2}{r^2}\frac{\partial v_r}{\partial \theta}\right], \tag{4.68}$$

this results simply in

$$\frac{\partial p}{\partial \theta} = 0. \tag{4.69}$$

Since we assume the flow to be fully developed, the velocity profile is identical at any axial position and symmetric about the axis of the pipe such that

$$\frac{\partial v_z}{\partial \theta} = 0. \tag{4.70}$$

When we now consider the z-component of momentum:

$$\frac{\partial v_z}{\partial t} + v_r\frac{\partial v_z}{\partial r} + \frac{v_\theta}{r}\frac{\partial v_z}{\partial \theta} + u\frac{\partial v_z}{\partial z} = -\frac{1}{\rho}\frac{dp}{dz} + v\left[\frac{1}{r}\frac{\partial}{\partial r}\left(r\frac{\partial v_z}{\partial r}\right) + \frac{1}{r^2}\frac{\partial^2 v_z}{\partial \theta^2} + \frac{\partial^2 v_z}{\partial z^2}\right],$$
$$\tag{4.71}$$

we find that nearly every term vanishes, and thus we obtain the following:

$$\frac{1}{r}\frac{\partial}{\partial r}\left(r\frac{\partial v_z}{\partial r}\right) = \frac{1}{\mu}\frac{dp}{dz}, \tag{4.72}$$

where $p = p(z)$ and $v_z = v_z(r)$. Note that this result is similar to the expression written for steady duct flow in Cartesian coordinates. Here, we continue a step further and solve for the velocity profile in the round pipe.

Since the pressure gradient $\left(\frac{dp}{dz}\right)$ is independent of radius r, we can integrate twice:

$$r\frac{\partial v_z}{\partial r} = \int \frac{1}{\mu}\frac{dp}{dz}r\,dr$$

$$= \frac{1}{2\mu}\frac{dp}{dz}r^2 + C_1 \tag{4.73}$$

and therefore:

$$v_z(r) = \int \left(\frac{1}{2\mu}\frac{dp}{dz}\frac{r^2}{r} + \frac{C_1}{r}\right) dr$$

$$= \frac{1}{4\mu}\frac{dp}{dz}r^2 + C_1 \log(r) + C_2, \tag{4.74}$$

where C_1 and C_2 are the constants.

Now applying the following boundary conditions:

1. At the centerline: $v_z(r=0) \neq 0$. We do not know the centerline velocity, but, of course, it must be physical. Furthermore, $\log(0) \to \infty$ such that $C_1 = 0$; and
2. At the pipe wall $v_z(r = R) = 0$ in order to satisfy the no-slip condition. Therefore, we obtain

$$C_2 = -\frac{1}{4\mu}\frac{dp}{dz}R^2. \tag{4.75}$$

Plugging the constants $(C_{1,2})$ into Eq. (4.74), we can now simplify further and obtain the exact velocity field for Hagen–Poiseuille flow:

$$v_z(r) = -\frac{1}{4\mu}\frac{dp}{dz}(R^2 - r^2) \tag{4.76}$$

$$v_\theta = v_r = 0.$$

This velocity distribution represents a fully developed, laminar pipe flow in the form of a paraboloid with revolution about the centerline. The velocity variation then depends only on an external pressure drop.

It is curious to note here that Poiseuille, a French physicist and physiologist, had an interest in human blood flow, whereas Hagen, a German civil engineer, focused on hydraulics. Each, independently, observed an empirical relationship between pressure drop and pipe diameter for laminar conditions with constant flow rate:

$$Q = \int_{cs} v_z\,dA, \tag{4.77}$$

where cs represents the cross-sectional control surface of the pipe and $dA = 2\pi r dr$.
Using Eq. (4.76), we can then solve where

$$
\begin{aligned}
Q &= \int_{cs} v_z dA \\
&= 2\pi \int_0^R -\frac{1}{4\mu}\frac{dp}{dz}(R^2 - r^2)r dr \\
&= 2\pi \left[-\frac{1}{4\mu}\frac{dp}{dz}\left(\frac{R^2 r^2}{2} - \frac{r^4}{4} \right) \right]\Big|_0^R .
\end{aligned}
$$

(4.78)

Finally, evaluating across the radius of the pipe, we obtain

$$
Q = \frac{\pi R^4}{8\mu}\left(-\frac{dp}{dz} \right),
$$

(4.79)

which over a known length of pipe L, the pressure gradient can be determined as

$$
\Delta p = \frac{8\mu L Q}{\pi R^4},
$$

(4.80)

such that

$$
-\frac{dp}{dz} = \frac{\Delta p}{L}.
$$

(4.81)

For our parabolic velocity profile, the maximum velocity is

$$
v_{z,max} = v_z(r = 0) = \frac{R^2}{4\mu}\left(-\frac{dp}{dz} \right),
$$

(4.82)

and, as shown in Fig. 4.10, the average velocity can be defined as

$$
\begin{aligned}
\bar{v}_z = \frac{Q}{A} &= \frac{1}{\pi R^2} \cdot \frac{\pi R^4}{8\mu}\left(-\frac{dp}{dz} \right) \\
&= \frac{R^2}{8\mu}\left(-\frac{dp}{dz} \right) \\
&= \frac{1}{2}v_{z,max}.
\end{aligned}
$$

(4.83)

Since we are dealing with a fully developed flow, the energy level between any
two cross-sections along the pipe must be described through Bernoulli's equation,

Fig. 4.10 Maximum and mean velocities of a parabolic Hagen–Poiseuille flow

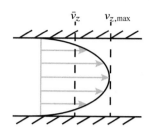

Fig. 4.11 Schematic representation of the energy balance inside an angled, straight pipe with pressure-driven flow

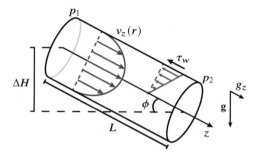

but with losses:

$$\left(H + \frac{p}{\rho g} + \frac{V^2}{2g}\right)_1 = \left(H + \frac{p}{\rho g} + \frac{V^2}{2g}\right)_2 - \Delta h_p + \Delta h_m + \Delta h_f, \qquad (4.84)$$

where Δh_p, Δh_m, and Δh_f represent *head* added through, for instance, a pump, minor losses associated with bends, valves and other obstructions, and frictional losses, respectively. It should be noted that $\Delta h_f \propto \tau_w$ (shear stress at the pipe wall). Finally, H is the elevation, above a reference plane, along the direction that gravitational acceleration g acts.

For the simplest case, as depicted in Fig. 4.11, $\Delta h_p = 0$ and $\Delta h_m = 0$. Thus, considering that mass must be conserved from one cross-section to the next ($V_1 = V_2$), the Bernoulli's equation with losses reduces to

$$\Delta h_f = \Delta H + \frac{\Delta p}{\rho g}. \qquad (4.85)$$

When we balance forces along the z-axis, we obtain

$$\sum F_z = \Delta p(\pi R^2) + \rho g(\pi R^2)L \sin(\phi) - \tau_w(2\pi R)L = \dot{m}(V_2 - V_1) = 0, \qquad (4.86)$$

which then reduces to

$$\frac{\Delta p}{\rho g} + \Delta H - \frac{2\tau_w L}{\rho g R} = 0, \qquad (4.87)$$

where $\Delta H = L \sin(\phi)$. In turn we can show that

$$\Delta h_f = \frac{4\tau_w}{\rho g} \frac{L}{D}, \tag{4.88}$$

where τ_w is generally unknown for pipe-flow problems, and therefore, empirical data must be used, together with this above result, in order to propose the following correlation:

$$\Delta h_f = f \frac{L}{D} \frac{V^2}{2g}, \tag{4.89}$$

where f is known as the Darcy friction factor and is dependent on cross-sectional shape, Re, and $\frac{\varepsilon}{D}$, where ε is the mean wall roughness. Values for f are easily extracted from an empirical database shown conveniently in the form of a Moody diagram (Moody 1944).

We then bring together Eqs. (4.88) and (4.89):

$$\frac{4\tau_w}{\rho g} \frac{L}{D} = f \frac{L}{D} \frac{V^2}{2g} \tag{4.90}$$

and obtain

$$f = \frac{8\tau_w}{\rho \bar{v}_z^2}, \tag{4.91}$$

where, for convenience, we now denote $V = \bar{v}_z$.

The wall shear stress, as shown in Fig. 4.12, can be related to deformation:

$$\tau_w \propto \frac{d\theta}{dt}, \tag{4.92}$$

where

$$\tan d\theta = \frac{dv_z dt}{dr} \tag{4.93}$$

and at the limit $\tan d\theta \rightarrow d\theta$.

Fig. 4.12 Description of wall shear stress (τ) in pipe flow

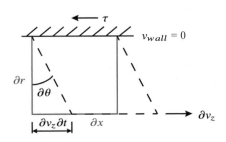

Thus, we can relate

$$\frac{d\theta}{dt} = \frac{dv_z}{dr} \propto \tau_w,$$

(4.94)

where μ is the constant of proportionality such that:

$$\tau_w = \mu \left(-\frac{dv_z}{dr} \right)_{r=R}.$$

(4.95)

For the case of a laminar flow (with Newtonian fluid):

$$\tau_w = -\mu \frac{d}{dr} \left(-\frac{1}{4\mu} \frac{dp}{dz} (R^2 - r^2) \right)_{r=R}$$

$$= \frac{R}{2} \left(-\frac{dp}{dz} \right).$$

(4.96)

Returning to Eq. (4.83)

$$-\frac{dp}{dz} = \frac{8\mu \bar{v}_z}{R^2},$$

(4.97)

the wall shear stress can be written as:

$$\tau_w = \frac{4\mu \bar{v}_z}{R}.$$

(4.98)

Finally, we can also relate the friction factor simply as

$$f = \frac{8}{\rho \bar{v}_z^2} \left(\frac{4\mu \bar{v}_z}{R} \right) = \frac{8}{\rho \bar{v}_z^2} \left(\frac{8\mu \bar{v}_z}{D} \right) = \frac{64}{Re_D}.$$

(4.99)

It is important to remember that the four previous equations apply only to laminar flow of a Newtonian fluid (Hagen–Poiseuille flow).

4.4.1 Exercise Comparing Flow Through Blood Vessels

Consider the flow of blood as it moves away from the heart through increasingly smaller vessels. Assuming blood has a viscosity of $\mu = 0.0035\,\text{NS/m}^2$ and density $\rho = 1060\,\text{kg/m}^3$ (empirical values), estimate the pressure gradient (dp/dz) and wall shear stress (τ_w) in the aorta, an arteriole, and a capillary. Table 4.1 contains measured values for each vessel.

Table 4.1 Documented values for blood vessels

Vessel	Diameter	Wall roughness (ϵ)	Mean velocity ($\bar{v}_z$)
Aorta	25 mm	1 μm	48 cm/s
Arteriole	0.5 mm	1 μm	3 cm/s
Capillary	0.008 mm	Smooth	0.7 cm/s

Recall relationships, derived in the previous section, between f, $\dfrac{dp}{dz}$ and τ_w for flow through a round pipe:

$$f = \frac{8\tau_w}{\rho \bar{v}_z^2},$$

$$(4.100)$$

and from Bernoulli's equation:

$$\frac{\Delta p}{\rho g} + \Delta z = -\frac{2\tau_w L}{\rho g R}$$

$$\frac{dp}{dz} = -\frac{4\tau_w}{D},$$

$$(4.101)$$

where we ignore any height differences ($\Delta z = 0$) and recall that the pressure drop is linear in a fully developed flow.

Beginning with the **aorta**, we calculate the Reynolds number:

$$\mathrm{Re}_D = \frac{\rho \bar{v}_z D}{\mu} = \frac{1060 \times 48 \times 10^{-2} \times 25 \times 10^{-3}}{0.0035} = 3634.$$

$$(4.102)$$

For this turbulent flow, we can look up $f = 0.042$ from the Moody diagram (Moody 1944) using Re_D and ϵ/D. Rearranging Eq. (4.100), the wall shear stress can be calculated such that:

$$\tau_w = \frac{f}{8} \rho \bar{v}_z^2 = \frac{0.042}{8} \times 1060 \times 0.48^2$$

$$= 1.3 \text{ Pa},$$

$$(4.103)$$

and the pressure gradient follows

$$\frac{dp}{dz} = -\frac{4\tau_w}{D} = \frac{4(1.3)}{25 \times 10^{-3}}$$

$$= -208 \text{ Pa/m}.$$

$$(4.104)$$

Tips

Recognize that we have made many assumptions and simplifications from real vessels when using the Moody diagram and the equations herein, including:

- The length of vessel is straight and has no bifurcations, valves, etc;
- The vessel walls are stiff (rigid);
- The flow is fully developed; and
- We assume blood to act as a Newtonian fluid.

Next, for the **arteriole**, flow is laminar:

$$\mathrm{Re}_D = \frac{1060 \times 3 \times 10^{-2} \times 0.5 \times 10^{-3}}{0.0035} = 4.5. \tag{4.105}$$

Again, assuming blood acts as a Newtonian fluid, we can use the equations we derived for Hagen–Poiseuille flow to estimate the following:

$$f = \frac{64}{\mathrm{Re}_D} = 14.2, \tag{4.106}$$

$$\tau_w = \frac{4\mu\bar{v}_z}{R} = 1.7\,\mathrm{Pa}, \text{ and} \tag{4.107}$$

$$\frac{dp}{dz} = -\frac{8\mu\bar{v}_z}{R^2} = -13.4\,\mathrm{kPa/m}. \tag{4.108}$$

Note that we could have used Eqs. (4.100) and (4.101) to reach approximately the same values.

Finally, flow through the **capillary** has a Reynolds number of $\mathrm{Re}_D = 0.017$, which would suggest Stokes flow. Following the same procedure as for laminar flow through the arteriole will have some inherent error but will allow us to approximate relative scales for f, τ_w, and dp/dz:

$$f = \frac{64}{\mathrm{Re}_D} = 3765, \tag{4.109}$$

$$\tau_w = \frac{f}{8}\rho\bar{v}_z^2 \approx \frac{4\mu\bar{v}_z}{R} = 24.4\,\mathrm{Pa}, \text{ and} \tag{4.110}$$

$$\frac{dp}{dz} = -\frac{4\tau_w}{D} \approx -\frac{8\mu\bar{v}_z}{R^2} = -12.2\,\mathrm{MPa/m}. \tag{4.111}$$

The shear stress and pressure gradient estimates have increased drastically this far from the heart. However, we have assumed blood is a single-phase liquid despite the vessel diameter is now on the same scale as a red blood cell. Cells are literally squeezing through the capillaries so we should expect a large wall shear stress!

Additionally, we see an increase in the pressure gradient as we branch into smaller vessels. As such blood in our large arteries moves much faster despite a smaller pressure drop.

4.5 Pipe Flow with Oscillating Pressure Gradient

Returning to the assumptions made for Hagen–Poiseuille flow in Sect. 4.4, we now wish to analyze the time-varying velocity profile in our pipe when the steady-state restriction is dropped. Specifically, this section will focus on pipe flow driven by an oscillating pressure gradient with the following assumptions (all of which also applied to steady Hagen–Poiseuille flow):

- Newtonian fluid
- No-slip at wall
- Laminar flow

- Fully developed
- Circular cross-section
- Rigid walls

Recalling the normalized Navier–Stokes equation presented in Sect. 3.1:

$$\text{St}\frac{\partial \mathbf{v}^*}{\partial t^*} + (\mathbf{v}^* \cdot \nabla^*)\,\mathbf{v}^* = \frac{1}{\text{Fr}^2}\mathbf{g}^* - \text{Eu}\nabla^* p^* + \frac{1}{\text{Re}}\nabla^{*2}\mathbf{v}^*, \tag{3.4}$$

we can no longer ignore the unsteady $\frac{\partial \mathbf{v}^*}{\partial t^*}$ term, and $\nabla^* p^*$ is now a time-varying quantity.

Following similar steps to the steady flow derivation, we can reduce the axial (z) momentum equation as follows:

$$\frac{\partial v_z}{\partial t} + \cancel{v_r\frac{\partial v_z}{\partial r}} + \cancel{\frac{v_\theta}{r}\frac{\partial v_z}{\partial \theta}} + \cancel{v_z\frac{\partial v_z}{\partial z}} = -\frac{1}{\rho}\frac{dp}{dz} + \upsilon\left(\frac{1}{r}\frac{\partial}{\partial r}(r\frac{\partial v_z}{\partial r}) + \cancel{\frac{1}{r^2}\frac{\partial^2 v_z}{\partial \theta^2}} + \cancel{\frac{\partial^2 v_z}{\partial z^2}}\right), \tag{4.112}$$

such that:

$$\frac{\partial v_z}{\partial t} = -\frac{1}{\rho}\frac{dp}{dz} + \upsilon\left(\frac{\partial^2 v_z}{\partial r^2} + \frac{1}{r}\frac{\partial v_z}{\partial r}\right). \tag{4.113}$$

We now assume that the pressure gradient varies sinusoidally with time:

$$\frac{dp}{dz} = -\rho K e^{i\omega t}, \tag{4.114}$$

where $\omega = 2\pi f$; $e_{i\omega t} = \cos\omega t + i\sin\omega t$; and $i = \sqrt{-1}$. The sinusoid has amplitude K and frequency f.

While its derivation (Womersley 1955) is beyond the scope of this textbook, the velocity profile of the long-term steady oscillation generated by dp/dz can be described by

$$v_z(r, t) = \frac{K}{i\omega} e^{i\omega t} \left[1 - \frac{J_0(r\sqrt{-i\omega/v})}{J_0(R\sqrt{-i\omega/v})} \right],$$ (4.115)

where J_0 is a Bessel function of the first kind and of zeroth order.

We can capture the general effect of the Bessel function using two overlapping series approximations:

$$\text{Small } x < 2 \quad \rightarrow \quad J_0(x) \approx 1 - \frac{x^2}{4} + \frac{x^4}{64} - \ldots$$

$$\text{Large } x > 2 \quad \rightarrow \quad J_0(x) \approx \sqrt{\frac{2}{\pi x}} \cos\left(x - \frac{\pi}{4}\right),$$ (4.116)

where $x \propto \omega$.

From Eq. (4.115), we can pick the following dimensionless variables:

$$r^* = \frac{r}{R}, \quad \omega^* = \frac{\omega R^2}{v} \quad \text{and} \quad v_z^* = \frac{v_z}{V_{max}},$$

where ω^* can be thought of as a kinetic Reynolds number, and $V_{max} = KR^2/4v$ is the centerline velocity for steady Poiseuille flow with a pressure gradient $-\rho K$.

When we insert Eq. (4.116) into Eq. (4.115), we obtain two series approximations for the real part of the velocity:

$$\text{Small } \omega^* < 4 \rightarrow \frac{v_z}{V_{max}} \approx (1 - r^{*2})\cos\omega t + \frac{\omega^*}{16}(r^{*4} + 4r^{*2} - 5)\sin\omega t + \mathcal{O}(\omega^{*2}),$$

$$\text{Large } \omega^* > 4 \rightarrow \frac{v_z}{V_{max}} \approx +\frac{4}{\omega^*}\left[\sin\omega t - \frac{e^{-B}}{\sqrt{r^*}}\sin(\omega t - B) \right] + \mathcal{O}(\omega^{*-2}),$$ (4.117)

where $B = (1 - r^*)\sqrt{\frac{\omega^*}{2}}$.

Recall that $\frac{dp}{dz} \propto \cos \omega t$ so, for small ω^*, velocity is nearly a quasi-steady Hagen–Poiseuille profile that oscillates with the pressure gradient. However, at large ω^*, velocity lags behind the pressure gradient by approximately 90°, and the centerline velocity is less than V_{max}.

In the medical community, oscillating (and pulsatile) velocity profiles are characterized using the dimensionless Womersley (1955) number α:

$$\alpha = R\sqrt{\frac{\omega}{v}},$$ (4.118)

which can be factored into Eq. (4.115) such that:

$$v_z(r, t) = \frac{K}{i\omega} e^{i\omega t} \left[1 - \frac{J_0(i^{3/2}\alpha r/R)}{J_0(i^{3/2}\alpha)} \right]. \tag{4.119}$$

Returning to the dimensionless groups extracted from the normalized Navier–Stokes equation, we realize the Womersley number is the product of two of those terms:

$$St_D Re_D = \frac{fD}{U_\infty} \frac{U_\infty D}{v} = \frac{4R^2 f}{v} = \frac{2}{\pi}\omega^* = \frac{2}{\pi}\alpha^2,$$

where St_D is the ratio of unsteady (or oscillatory) forcing to convective (inertial) forcing, and Re_D is the ratio of inertial to viscous effects.

The Womersley number, as defined in Eq. (4.118), is the ratio of transient inertial forcing from an oscillatory flow to viscous forcing. Alternatively, this can be thought of as the relation between the characteristic time associated with bringing the flow to a stop through friction:

$$t_{viscous} \propto \frac{\rho R^2}{\mu}, \tag{4.120}$$

and the characteristic time of the oscillation or pulsation:

$$t_{pulse} = \frac{1}{f} = \frac{2\pi}{\omega}. \tag{4.121}$$

Combining the above yields the Womersley number:

$$\alpha = \sqrt{\frac{2\pi t_{viscous}}{t_{pulse}}} = R\sqrt{\frac{\rho\omega}{\mu}} = R\sqrt{\frac{\omega}{v}}. \tag{4.122}$$

Changing α in Eq. (4.119) has a significant effect on the velocity profile shape, ranging from a near parabola at low α to plug-like flow for high α. Figure 4.13 shows velocity profiles at three different α and at various phases of the oscillation.

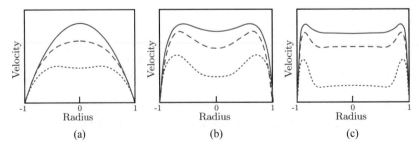

Fig. 4.13 Oscillatory velocity profile shape as a function of increasing Womersley number (α). The curves of each plot represent the profile at different phase angles during the oscillation. (**a**) $\alpha = 1$. (**b**) $\alpha = 8$. $\alpha = 15$

> **Tips**
>
> - Small α: friction dominates oscillatory effects, and the velocity profile is nearly parabolic;
> - Increasing α: fluid inertia becomes more dominant. The velocity profile flattens out in the core, there is flow reversal close to the walls, and the peak velocity no longer occurs along the pipe's centerline;
> - Large α: oscillating flow completely dominates, and the velocity profile is almost flat.
>
> Each of these effects can be visualized in Fig. 4.13 and the following exercise.

4.5.1 Exercise on Pulsatile Blood Flow

Recall that the equation for the velocity profile in a pipe due to an oscillating pressure gradient, $\frac{dp}{dx} = A^* e^{i\omega t}$, is given by the *real* part of the following expression:

$$v_z(r, t) = \frac{A^*/\rho}{i\omega} e^{i\omega t} \left[1 - \frac{J_0(i^{3/2}\alpha \, r/R)}{J_0(i^{3/2}\alpha)} \right], \tag{4.123}$$

where J_0 denotes a Bessel function of the first kind (order zero), α is the Womersley number, ω is the angular frequency, R is the radius of the pipe, and A^* is the amplitude of the pressure gradient.

A pulsatile flow can be mathematically represented by the superposition of the above oscillatory velocity profile and a steady component described by Hagen–Poiseuille flow, such that there is net flow in one direction.

Consider the flow of blood some 60 cm from the heart, through the aorta, toward the femoral arteries. Over this distance, the pressure drop due to the steady component of the flow is small: $\Delta p = 10$ Pa. For a heart rate of 70 beats per minute, the oscillatory component has $A^* = 7500$ Pa/m. Assuming the aorta has a diameter of 1 cm, and blood has a dynamic viscosity of 0.0035 Ns/m^2 and a density of 1060 kg/m^3, we can calculate and plot the pulsatile velocity profile in the aorta (60 cm from the heart).

We begin by computing the Womersley number using the heart rate:

$$\alpha = \frac{D}{2} \sqrt{\frac{2\pi \rho f}{\mu}} = \frac{0.01}{2} \sqrt{\frac{2\pi \times 1060 \times (70/60)}{0.0035}} = 7.5. \tag{4.124}$$

Using α and the real part of Eq. (4.123), the oscillatory velocity profile can be calculated (as plotted in Fig. 4.14).

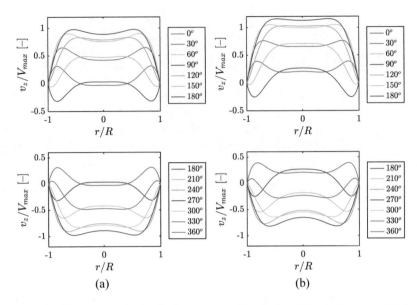

Fig. 4.14 Oscillatory (**a**) and pulsatile (**b**) velocity profiles for blood flow in the aorta at 70 beats per minute ($\alpha = 7.5$)

The steady component of the pulsatile flow can be represented by a pressure-driven pipe flow (Hagen–Poiseuille) where the pressure gradient is approximated as a pressure drop (Δp) over the distance from the heart ($L = 60$ cm), such that:

$$v_{z,\text{steady}}(r) = -\frac{1}{4\mu}\frac{\Delta p}{L}(R^2 - r^2), \qquad (4.125)$$

where $\Delta p = -10$ Pa.

Adding the steady flow parabola to our oscillatory component produces the pulsatile velocity profile shown in Fig. 4.14, where the velocity has been normalized by the peak velocity of the oscillatory flow. The influence of the steady flow is, obviously, most apparent along the centerline, but the velocity magnitude is increased throughout.

Now, let us consider what occurs in the flow when the heart rate increases or decreases (e.g., during heavy exercise or due to damage to heart tissue). Figure 4.15 demonstrates the influence α has on the velocity profile of blood flow in the aorta. At $\alpha = 3$, the inertial pulse period is long and frictional forces are allowed to dominate. The profile is nearly parabolic for the majority of the pulse. At elevated heart rates ($\alpha = 15$), the fluid inertia overpowers viscous effects and the velocity profile would be nearly flat across the width of the vessel if not for the steady component, which maintains a net positive flow rate. Frictional effects are only noticeable close to the walls, where velocity fluctuations are the largest.

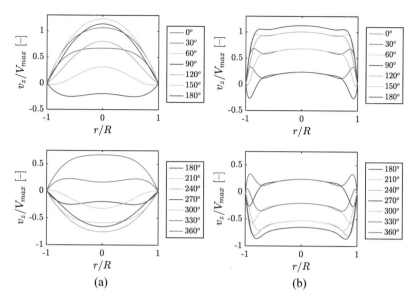

Fig. 4.15 Pulsatile velocity profiles for blood flow in the aorta at $\alpha = 3$ (**a**) and $\alpha = 15$ (**b**)

Tips

Note that while this exercise assumed a simple sinusoidal pressure gradient, represented by the sum of an oscillatory and steady component using Eqs. (4.123) and (4.76), respectively, more complex pressure signals, like the cardiac waveform, can be reduced to the sum of sinusoidal harmonics using Fourier analysis. Each of these harmonics has a unique α and can be modeled using Eq. (4.123).

4.6 Pulsatile Flow in Flexible Pipes

We now leave the rigid wall assumption behind and focus our attention on pulsatile flows through flexible pipes that are more common in the biological world. Unlike our work in the previous section on pulsatile duct flow, we must now take into account the influence of wave speed, i.e., pressure and flow profiles are no longer synchronous.

In dealing with the mammalian cardiovascular system, which is comprised of non-linearly elastic materials, we first must recall Hooke's law:

$$\sigma = E\epsilon, \tag{4.126}$$

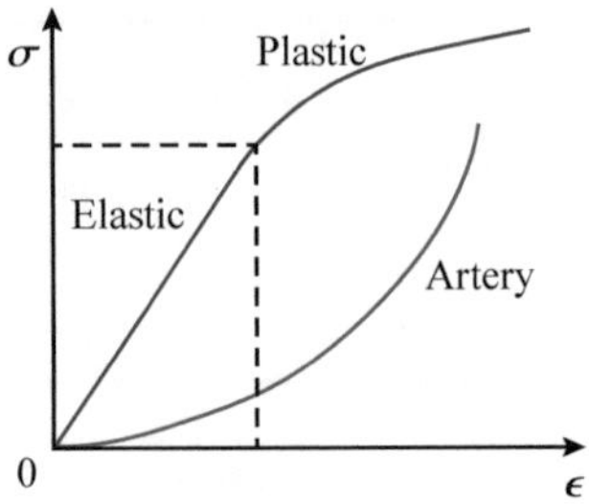

Fig. 4.16 Stress–strain curves for three common material types

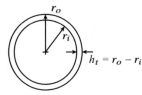

Fig. 4.17 A thin-walled pressure vessel has a wall thickness that is 10–15% of its inner radius

where E is the modulus of elasticity (units N/m^2). Figure 4.16 shows the stress–strain relationship for an elastic, a plastic, and a non-linearly elastic material (e.g., an artery). Stress in a non-linear material can be represented by the integral of incremental changes in the elasticity (E_{inc}) due to increasing strain:

$$\sigma_{\text{non-linear}} = \int_0^{\epsilon_n} E_{inc} d\epsilon. \tag{4.127}$$

Arteries are complex structures composed of multiple layers with two primary load-bearing fibers: elastin and collagen. The proportion and arrangement of each significantly affects the mechanical response of the vessel. Arteries are most accurately described as viscoelastic, implying that, in addition to behaving non-linearly, they are also dependent on the rate of strain:

$$\sigma_{\text{viscoelastic}} = E_1 \epsilon + E_2 \frac{d\epsilon}{dt}. \tag{4.128}$$

Additionally, arteries are metabolically active materials and can contract and expend energy to resist strain.

To estimate the *compliance* of an artery, we can consider it as a thin-walled pressure vessel where its wall thickness ($h_t = r_o - r_i$) is

$$0.1 \leq \frac{h_t}{r_i} \leq 0.15, \tag{4.129}$$

and r_i and r_o are the inner and outer radii, respectively (as depicted in Fig. 4.17). The compliance of the vessel can be determined as a way of relating circumferential hoop stress (S_h) to a change in the vessel's cross-sectional area.

Fig. 4.18 Balance of forces acting on a circular thin-walled vessel due to a transmural pressure (per unit length)

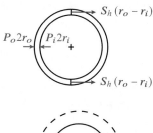

Fig. 4.19 Incremental hoop strain results in small radial expansions (dr)

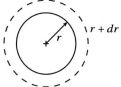

We begin by considering the balance of forces acting on the walls of the vessel (see Fig. 4.18):

$$p_o 2r_o + 2S_h(r_o - r_i) = p_i 2r_i,$$

$$S_h h_t = p_i r_i - p_o r_o, \tag{4.130}$$

$$S_h \approx \frac{pr}{h_t},$$

where P is the transmural pressure and for thin walls $r \approx r_i \approx r_o$.

We recall that the incremental hoop strain ($d\epsilon_h$) can be related to the change in radius depicted in Fig. 4.19:

$$d\epsilon_h = \frac{dr}{r}. \tag{4.131}$$

For a thin-walled pressure vessel, the modulus of elasticity can be related to the hoop stress as follows:

$$E = \frac{dS_h}{d\epsilon_h}, \tag{4.132}$$

such that, with Eqs. (4.131) and (4.132), we can write

$$dS_h = E\frac{dr}{r}. \tag{4.133}$$

We can now rewrite Eq. (4.130) for an incremental stress:

$$dS_h = d\left(\frac{pr^2}{r_o h_o}\right), \tag{4.134}$$

where $r_o h_o \approx r h$ for an incompressible vessel of fixed length. The right-hand side can be expanded to

$$d(pr^2) = 2prdr + r^2 dp, \tag{4.135}$$

where $dr \approx 0$ for small strains.

Combining Eqs. (4.133) and (4.134), we obtain

$$E\frac{dr}{r} = \frac{r^2 dp}{r_o h_o}, \tag{4.136}$$

and we can now solve for the change in (transmural) pressure as a function of radius (or cross-sectional area).

Rearranging and integrating Eq. (4.136):

$$\int_{p_o}^{p} dP = E h_o r_o \int_{r_o}^{r} \frac{dr}{r^3}$$

$$p - p_o = \frac{E h_o r_o}{2}\left[\frac{1}{r_o^2} - \frac{1}{r^2}\right] \tag{4.137}$$

$$\text{or} \quad p - p_o = \frac{E h_o}{2 r_o}\left[1 - \frac{A_o}{A}\right].$$

And now, solving for area ratio:

$$\frac{A}{A_o} = \left[1 - \frac{(p - p_o) 2 r_o}{E h_o}\right]^{-1}. \tag{4.138}$$

The above equation can be written in the form of a polynomial series, where $C_1 = \frac{2 r_o}{E h_o}$, such that:

$$\frac{A}{A_o} = 1 + C_1(p - p_o) + [C_1(p - p_o)]^2 + [C_1(p - p_o)]^3 + \cdots \tag{4.139}$$

Multiplying through by A_o and then taking the derivative with respect to pressure produces

$$\frac{dA}{dp} = A_o C_1 + A_o C_1^2(p - p_o) + A_o C_1^3(p - p_o)^2 + \cdots, \tag{4.140}$$

which now directly relates a change in cross-sectional area to a change in transmural pressure.

However, C_1^2, C_1^3, etc. are small, such that we obtain an approximate expression for *compliance* (C):

$$C = \frac{dA}{dp} \approx A_o C_1 = \frac{2\pi r_o^3}{E h_o},$$ (4.141)

from which the modulus of elasticity (using a thin-wall assumption) can be estimated:

$$E = \frac{2\pi r_o^3}{C h_o}.$$ (4.142)

4.6.1   Exercise on Pressure Variation and Compliance in the Aortic Arch

In this exercise, we wish to produce a model for the time-dependent pressure fluctuations in the human aorta at a distance $L = 200$ mm from the heart. The model must account for the compliance of the vessel. Figure 4.20 shows a schematic representation of the problem, as well as important quantities that were recorded during strenuous activity (i.e., elevated heart rate).

Fig. 4.20 Schematic diagram and test subject data

Quantity	Value	Units
p_0	150	Pa
p_s	7500	Pa
n	2	Hz
D_0	20	mm
E	13700ϵ	Pa
L	200	mm
h_0	1	mm
μ_{blood}	0.0035	Ns/m^2
ρ_{blood}	1060	kg/m^3

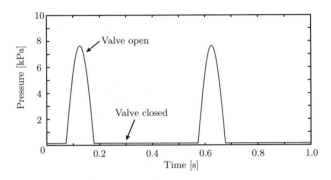

Fig. 4.21 Simple model of pressure variation at the heart

To begin, we can approximate the pressure signal at the heart ($p_1(t)$) assuming the pressure rises and falls quickly when the aortic valve is open and is otherwise constant, as illustrated in Fig. 4.21. Each heart beat consists of a single valve actuation. The described pressure signal can be modeled using the following sinusoidal function:

$$p_1(t) = p_0 + 2p_s \left(\sin^3(2\pi tn) - \frac{1}{2} \right), \tag{4.143}$$

where p_s is the static pressure, p_0 is the pressure far downstream, n is the heart rate, and the inequality prevents retrograde flow (ignored in this exercise for simplicity).

Now that we know the blood pressure during ejection from the heart, we require a model for p_2 at a distance L along the flexible aorta. To relate p_2 to the vessel compliance, we can define the pressure in terms of a strain, $\epsilon = (D - D_0)/D_0$, and a coefficient K similar to compliance:

$$p_2 = K \frac{D(t) - D_0}{D_0}, \tag{4.144}$$

where D is the time-dependent aorta diameter and D_0 is the initial, unstrained diameter.

To solve for the unknown K, recall:

$$\frac{dA}{dp} = \frac{\pi r_0^3}{E h_0}, \tag{4.145}$$

where h_0 is the arterial wall thickness, and $A = \frac{\pi D_0^2}{4}$ assuming a circular cross-section. The left-hand side then becomes

$$\frac{\pi D_0}{2} \cdot \frac{dD}{dp} = \frac{\pi r_0^3}{E h_0}, \tag{4.146}$$

such that:

$$\frac{dD}{dp} = \frac{D_0^2}{4Eh_0}.$$

(4.147)

Using Eq. (4.144) and $D - D_0 = dD$, we can solve for K:

$$K = \frac{D_0}{dD/dp} = D_0 \frac{4Eh_0}{D_0^2} = \frac{4Eh_0}{D_0}.$$

(4.148)

Before we can plot our function for aortic pressure, we must determine the rate of change of the diameter in response to the pressure wave. Consider:

$$\frac{dD}{dt} = \frac{2}{\pi L}(Q_1 - Q_2),$$

(4.149)

where the flow rates Q_1 and Q_2 are easily obtained using the Hagen–Poiseuille model for flow based on known pressure drops along the vessel, such that:

$$Q_1 = \frac{\pi D^4}{128\mu L}(p_1 - p_2)$$

$$Q_2 = \frac{\pi D^4}{128\mu L}(p_2 - p_0).$$

(4.150)

Substituting Q_1 and Q_2 into Eq. (4.149), we arrive at a function for the rate of change of the aortic diameter:

$$\frac{dD}{dt} = \frac{1}{64\mu L^2}D^4(p_0 + p_1 - 2p_2).$$

(4.151)

Using Eqs. (4.143), (4.144), and (4.151), the system of equations can be solved iteratively to produce the pressure signal downstream of the heart in a flexible vessel. Figure 4.22 shows p_1 and p_2 plotted over the course of two heart beats. The pressure in the aorta (p_2) has a significantly lower magnitude and is out of phase, compared to that in the left ventricle (p_1). The phase lag is a result of the compliance of the system: the expanding walls of the aorta slow down the transmission of the pressure wave generated by the contraction of the left ventricle. If the blood vessel were a rigid pipe, the phase lag would not be present. The pressure downstream is of lower amplitude due to the pressure gradient required to drive the flow against viscous effects.

Note that we have assumed an elastic relationship ($\sigma = E\epsilon$) for the vessel in this exercise, while in actuality our arteries are viscoelastic and respond more in the following form:

$$\sigma_{\text{viscoelastic}} = E_1\epsilon + E_2\frac{d\epsilon}{dt}.$$

(4.128)

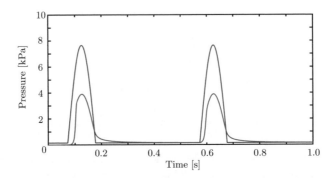

Fig. 4.22 Pressure fluctuations in aorta 200 mm downstream from the heart (red curve) as a result of the signal measured at the heart (blue)

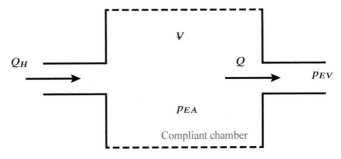

Fig. 4.23 The cardiovascular system is modeled as a simple Windkessel model where the compliant chamber (i.e., its volume V is variable) represents our elastic arteries downstream of the heart

4.7   Wave Propagation in Flexible Pipes

The previous exercise introduced the idea of a pressure wave originating from the heart and propagating through the aorta. Let us again consider a pressure pulse at the root of the aorta and compare the excess pressure in the aorta p_{EA} (effectively upstream of the circulatory system) to the excess pressure in the vena cava p_{EV} (on its return to the heart). Recall that for a rigid pipe system, p_{EA} and p_{EV} would be identical. When the system is flexible, we must account for its compliance.

The Windkessel effect, theorized by Frank (1899), describes arterial blood pressure relative to the compliance of elastic arteries and the resistance of small vessels in the cardiovascular system. A simple Windkessel model, where the heart and system of blood vessels are represented as a hydraulic loop with a compliant chamber, is shown in Fig. 4.23.

As a first, very crude approximation, we can relate the blood pressure to the flow rate Q using the (constant) compliance C and the volume of the Windkessel chamber:

$$p_{EA} = \frac{\mathcal{V}}{C}, \text{ and}$$

$$\frac{d\mathcal{V}}{dt} = Q_H - Q, \tag{4.152}$$

where Q_H is the blood flow rate leaving the heart.

However, a Windkessel model is insufficient to describe the observed pressure pulse (and its reflections as it encounters bifurcations) that travels through the circulatory system as a wave.

Instead, we consider the water-hammer effect, which describes the generation of a pressure wave when a fluid undergoes a sudden change in momentum. This effect is common in flexible systems but is even observed in fully rigid pipes when a valve is suddenly closed. How a system responds to such a change is described below.

When a valve is closed, a force must be applied to decelerate all of the fluid in the pipe, which can be represented by the balance between the pressure force and the fluid deceleration:

$$\Delta p \frac{\pi D^2}{4} = \rho L \frac{\pi D^2}{4} \frac{dU}{dt}, \tag{4.153}$$

and for a valve closing over Δt:

$$\Delta p \approx \frac{\rho L U}{\Delta t}. \tag{4.154}$$

In practice, even rigid pipes will deform (L is allowed to change), and the fluid will compress slightly. For a short section of pipe (Δz):

$$\Delta p = -\rho \Delta z \left(\frac{-v_z}{\Delta t} \right), \tag{4.155}$$

where $-\frac{v_z}{\Delta t} = \frac{dU}{dt}$.

The relevant time change in this scenario, as depicted in Fig. 4.24, is the time for the upstream end of our short fluid element (defined by Δz) to come to a stop:

$$\Delta t = \frac{\Delta z - v_z \Delta t}{a}$$

$$= \frac{\Delta z}{a + v_z}, \tag{4.156}$$

where a is the pressure wave speed (i.e., the speed at which the disturbance propagates away from the valve). We can now rewrite Eq. (4.155)

$$\Delta p = -\rho \Delta z \left[\frac{-v_z}{\Delta z / (a + v_z)} \right]$$

$$= \rho v_z a \left(1 + \frac{v_z}{a} \right). \tag{4.157}$$

Fig. 4.24 When a valve
suddenly closes, a pressure
wave (red dashed line)
propagates back through the
pipe that brings the fluid to
rest and may cause the pipe
walls to expand

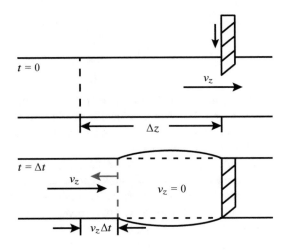

In stiff systems, like water flowing through a steel pipe, $a \gg u$ such that

$$\Delta p \approx \rho v_z a. \tag{4.158}$$

Let us now calculate the wave speed by evaluating the change in volume due to
fluid compression and pipe expansion:

$$
\begin{aligned}
\Delta V &= \frac{\pi D^2}{4} \Delta z \left(\frac{1}{k}\right) \Delta p + \frac{1}{E} \frac{\Delta p D}{2h_t} \frac{D}{2} \pi D \Delta z \\
&= \frac{\pi D^2}{4} \Delta z \left(\frac{1}{k} + \frac{D}{E h_t}\right) \Delta p,
\end{aligned}
\tag{4.159}
$$

where k is the fluid's bulk modulus, and E and h_t are the elastic modulus and wall
thickness of the pipe, respectively.

> **Tips**
> The water-hammer effect is potentially harmful to many fluidic systems. The
> surge of pressure when a valve closes causes:
>
> - Fluid compression as it is rapidly decelerated ($\frac{\pi D^2}{4} \frac{\Delta z \Delta p}{k}$ in Eq. (4.159));
>   and
> - Pipe expansion ($\frac{1}{E} \frac{\Delta p D}{2h} \frac{D}{2}$ accounts for the change in pipe diameter).

The pressure wave speed can be taken as the time needed for the flow to fill the
increased volume and then equated to Eq. (4.156):

$$\Delta t = \frac{\Delta V}{Q} = \frac{\frac{\pi D^2}{4} \Delta z \left(\frac{1}{k} + \frac{D}{Eh_t} \right) \Delta p}{\frac{\pi D^2}{4} v_z} = \frac{\Delta z}{v_z + a}. \tag{4.160}$$

Now, substitute Eq. (4.157) into Eq. (4.160), and one obtains the wave speed:

$$a = \left[\rho \left(1 + \frac{v_z}{a} \right)^2 \left(\frac{1}{k} + \frac{D}{Eh_t} \right) \right]^{-\frac{1}{2}}. \tag{4.161}$$

This represents the speed at which a pressure wave, generated by a reflection at a bifurcation or closed valve, propagates through the pipe. For stiff systems:

$$a \approx \frac{1}{\sqrt{\rho \left(\frac{1}{k} + \frac{D}{Eh_t} \right)}}, \tag{4.162}$$

and for liquids in flexible tubes (where $k \ll E$):

$$a \approx \sqrt{\frac{Eh_t}{D\rho}}. \tag{4.163}$$

Tips
In stark contrast with one another, wave speeds in industrial and biological systems can vary as follows:

- $1000 \lesssim a \lesssim 1500$ m/s for water in steel pipes; and
- Only $3 \lesssim a \lesssim 12$ m/s for blood in large arteries!

References

Couette, M. (1890). Études sur le frottement des liquides. *Annales de Chimie et de Physique, 20,* 433–510.

Einstein, A. (1906). Eine neue Bestimmung der Moleküldimensionen. *Annalen der Physik, 324*(2), 289–306.

Einstein, A. (1911). Bemerkung zu dem Gesetz von Eötvös. *Annalen der Physik, 339*(1), 165–169.

Frank, O. (1899). Die Grundform des Arteriellen Pulses. *Zeitschrift für Biologie, 37,* 483–526.

Hagen, G. H. L. (1839). Über die Bewegung des Wassers in engen zylindrischen Röhren. *Poggendorf's Annalen der Physik und Chemie, 46,* 423–442.

Jeffreys, H. (1927, 1931). *Operational Methods in mathematical physics.* Cambridge University Press.

Maxwell, J. C. (1867). On the dynamical theory of gases. *Philosophical Transactions of the Royal Society of London, 157,* 49–88.

Moody, L. (1944). Friction factors for pipe flow. *Transactions of the American Society Mechanical Engineers, 66*(8), 671–678.

Mooney, M. (1951). The viscosity of a concentrated suspension of spherical particles. *Journal of Colloid Science, 6*(2), 162–170.

Poiseuille, J. L. M. (1840). Recherches expérimentales sur le mouvement des liquides dans les tubes de très petits diamètres, I–III. *Comptes Rendus de l'Académie des Sciences, 11*, 961–967, 1041–1048.

Poiseuille, J. L. M. (1841). Recherches expérimentales sur le mouvement des liquides dans les tubes de très petits diamètres, IV. *Comptes Rendus de l'Académie des Sciences, 12*, 112–115.

Womersley, J. R. (1955). Method for the calculation of velocity, rate of flow and viscous drag in arteries when the pressure gradient is known. *The Journal of Physiology, 127*(3), 553–563.

Chapter 5
External Flows

The fields of aerodynamics and hydrodynamics are vast and include the effects of compressibility, as well as cavitation, free-surface effects, boundary-layer control, to name but a few examples. With classical applications such as aircraft and ships, many have been characterized by very high speeds (high Reynolds numbers), among other parameters like the Froude and Mach numbers. These topics are often treated with tools derived from potential-flow theory, experiment, or advanced numerical simulations, which are beyond the scope of this book. Instead, in this chapter, we introduce the fundamentals of incompressible aero- and hydrodynamics and pick a few salient topics and exercises with which to promote exploration of specific biological and bio-inspired problems, acknowledging the fact that we are only scratching the surface. If we limit ourselves to *solutions* observed in our current snapshot of evolutionary time, or focus only on improving the performance of existing, human-designed propulsion systems (that in some sense also suffer from their own form of phylogenetic inertia), we might end up missing complete swaths of the solution space. In that vein, this chapter seeks to promote alternative perspectives of classical problems all the while encouraging the development of our own unique questions.

5.1 Strip Theory

The world is three-dimensional. There is no argument there. Even an albatross, with its exceptionally broad wingspan (distance from tip-to-tip), will nevertheless exhibit bulk three-dimensional flow at its wingtips, and at the roots of its wings. However, in many cases, a *strip-theory* approach, one where we break down a three-dimensional world into a series of two-dimensional (infinite and independent) sections, is highly effective, if not at least practical. If we look at the pioneering observations from Lilienthal (1889), it is immediately clear how this approach to study a white stork

© The Author(s), under exclusive license to Springer Nature Switzerland AG 2022
D. E. Rival, *Biological and Bio-Inspired Fluid Dynamics*,
https://doi.org/10.1007/978-3-030-90271-1_5

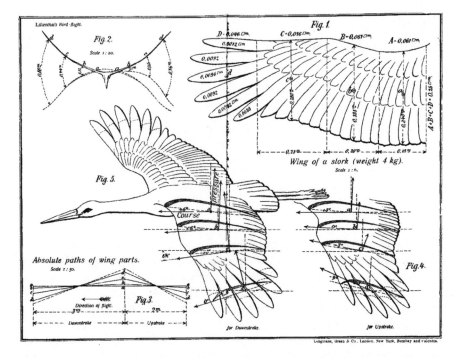

Fig. 5.1 In his analysis of bird flight as inspiration for aeronautical applications, Lilienthal (1889) divided a white stork's wing into two-dimensional strips each with their own local profile and angle of attack (as a function of wing stroke). At bottom both bottom- and top-stroke positions shown

(see Fig. 5.1) would be helpful later when designing aircraft, or any other moving body for that matter.

Accepting, first and foremost, that our objective is to start simple, and gradually build in complexity (i.e., three-dimensionality and unsteadiness), we will begin here by envisioning how all appendages or *propulsors* (e.g., wings, fins, flukes) can be broken down into two-dimensional pieces, as shown in Fig. 5.2. In such a manner, we can utilize known values corresponding to the two-dimensional (sectional) performance of specific shapes, which are in turn a function of parameters such as Reynolds number (i.e., length scale and fluid properties), and then simply integrate across the *strips* to obtain a bulk force or moment.

Now, at this point, we must generalize and assume that all appendages share common features, regardless of whether their role is to generate lift (L), thrust (T), drag (D), a pitching/rolling/yawing moment (M), or often as Nature prefers it, an ever-changing combination of all of these aforementioned forces and moments. Note that on a fin or fluke, a lift force acts to rotate and maneuver the body in a seemingly buoyant world. Unlike in engineering design, natural solutions address multiple functions with the same appendage. Inevitably, when designing an aircraft, lift and thrust generation are clearly distinct via wings and propellers/engines,

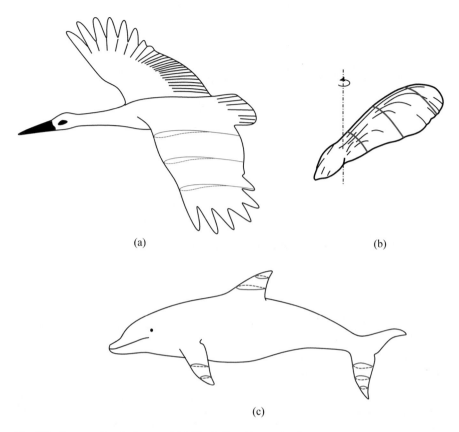

(a)   (b)

(c)

Fig. 5.2 Cross-sectional slices on (**a**) Lilienthal's white-stork wings, (**b**) an auto-rotating samara, as well as on (**c**) a dolphin's flippers and flukes. Strip theory provides us with a means to tackle a three-dimensional world into a series of infinite two-dimensional sections

respectively. We will return to this concept in due course through the exercise in Sect. 5.3.1.

In the spirit of abstraction, Fig. 5.3 shows us a generic two-dimensional section with all key jargon, which we often refer to as an airfoil. Of course, such a form can be used in other mediums such as water, and in such cases, we often refer to them as hydrofoils or, simply, foils. We should also avoid confounding the term airfoil with *wing*, as the latter implies a collection of two-dimensional sections (or strips). Furthermore, when considering the vast range of scales and materials, our abstracted foil should only be used to generalize concepts. Take the stork's wing, for instance. Whereas the leading-edge (LE) region is composed of bone and muscle, the bulk of the profile will be extremely thin and composed of feathers. Even thinner sections would be appropriate to describe bat or pterosaur membranes, and when looking at insect wings, the cross-sections may even appear corrugated and seemingly blunt at times. This variation, of course, is strongly dependent on Reynolds number, and as

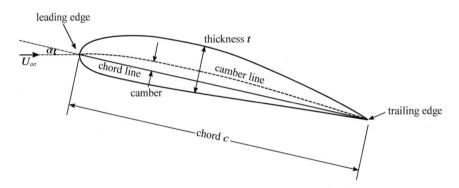

Fig. 5.3 Overview of key airfoil (hydrofoil) nomenclature including camber and chord lines. A camber line runs along the center between upper (suction) and lower (pressure) surfaces, whereas a chord line simply connects leading edge (LE) and trailing edge (TE) to each other with the shortest path (chord c). The angle of attack α represents the angle between the chord line and the incoming/relative freestream flow U_∞

Fig. 5.4 The pressure (C_p) and friction (C_f) coefficients demonstrating the local contributions to lift, moment, and drag over a generic airfoil, respectively. In this depiction we generalize and show that lift and pitching moment are dominated by pressure forces (left), whereas the drag is dominated by the wall shear-stress distribution (right). Of course these relative contributions are a strong function of Reynolds number, shape, angle of attack (α), and as we will see later, also unsteady (accelerating) conditions. Also denoted are the aerodynamic center (AC) and the center of gravity (CG)

we will see later, at lower Reynolds numbers ($10 < \mathrm{Re} < 10^3$), the flow separates at the leading edge and the resulting *leading-edge* vortices (LEVs) are used to generate lift and perform maneuvers.

We must also consider how classically we assume that streamlined appendages, like airfoils, result in drag through friction (integral of the wall shear stress τ along the surface) that can dominate over contributions from the pressure force, where we depict pressure and shear-stress forces in Fig. 5.4. Of course, this balance is a strong function of Reynolds number. Furthermore, appendages like wings may be used for generating lift at times, but must also be used to maneuver and even perch. Therefore, it is important to recognize that the same airfoil section may operate under differing conditions, i.e., attached and stalled conditions, and are not strictly optimal for only one function.

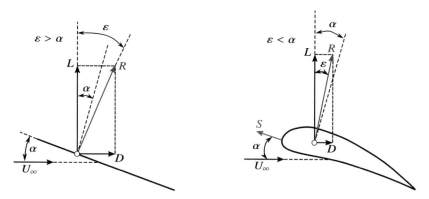

Fig. 5.5 On the left, the sharp leading edge promotes local separation and additional drag. In contrast, the rounded leading edge on the right allows for additional low pressure, which creates local suction, and therefore thrust (S), which in turn opposes and hence minimizes some of the body's drag (note how D is now much smaller)

In that vein, we must also discuss the concept of *leading-edge suction*, as shown in Fig. 5.5 for both sharp and rounded leading edges. Leading-edge suction is a key phenomenon only possible with our more traditional rounded leading-edge geometries and at reasonable Reynolds number (e.g., Re $\gtrsim 10^5$) for which the boundary layer does not separate; recall the topic of boundary-layer separation in Sect. 3.3. Here we see how a thin wing with a sharp leading edge—a simple abstraction of an insect wing—is unable to create this additional suction force (S). In Fig. 5.5, on the left-hand side, the flow will separate locally at the leading edge and additional drag will be produced. In contrast, for the right-hand side, the rounded leading edge will generate a low-pressure region (recall Bernoulli), in turn promoting a suction force directed upstream. This suction force acts to combat drag and tilt the resultant force vector (R) forward such that $\epsilon < \alpha$, as depicted for the case of leading-edge suction in Fig. 5.5.

In contrast, it is worth noting that a sharp trailing edge, on the other hand, is key to ensuring parallel flow as the two boundary layers meet in the wake. This constraint is known as the Kutta condition and will be key in later discussions.

Imagining all possible profile shapes, including many in nature that are also flexible and adaptive in time, we should define a few simple terms that we can best use to differentiate across a body of interest (see Fig. 5.6). We can imagine thick flukes versus extremely thin contours in the outer region of a bird's wing, formed by overlapping hand feathers. When generating lift, camber (curvature) assists in turning the flow via asymmetry. Conversely, an appendage used for control or turning, e.g., a fish's fin, best be symmetric to provide equal control in both directions.

A given profile's overall performance coefficients, characterizing force (or moment) as a function of unit span include lift (C_L), drag (C_D), and moment (C_M) coefficients, which are typically defined as

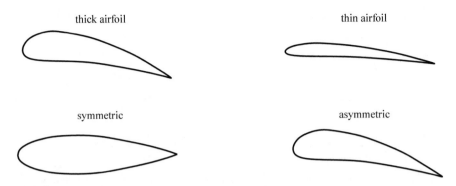

thick airfoil

thin airfoil

symmetric

asymmetric

Fig. 5.6 Although variation in appendage cross section is vast, we often choose to use these terms to categorize profiles and their respective performance. Of course, thin in the context of a feathered wing, or on a bat's membrane, is highly relative. We should also remind ourselves that most appendages in biological systems are soft and hence deformable, and possibly have an actively controlled shape

$$C_L = \frac{L}{\frac{1}{2}\rho U_\infty^2 c}; \tag{5.1}$$

$$C_D = \frac{D}{\frac{1}{2}\rho U_\infty^2 c}; \text{ and} \tag{5.2}$$

$$C_M = \frac{M}{\frac{1}{2}\rho U_\infty^2 c^2}, \tag{5.3}$$

where ρ is the fluid density, U_∞ represents the relative freestream velocity, and c is the local chord. It should be noted that C_L and C_D are defined as acting normal and parallel to the incoming fluid velocity (U_∞), whereas C_M must be defined relative to a chordwise position such that $C_M \neq 0$ when the center of pressure (aerodynamic center, AC) acts away from the airfoil's center of gravity (CG); see Fig. 5.4. For the purpose of describing typical aerodynamic (or hydrodynamic) features let us look at a generic, and relatively thin, symmetric NACA0009 airfoil ($t/c = 0.09$), as shown in Fig. 5.7. Acknowledging the fact that there are literally thousands of such empirical data sets available over a vast range of conditions, here we only look at this exemplary—and relatively boring!—thin and symmetric profile, observing basic features such as the onset of stall (boundary-layer separation), and the position of maximum L/D in the *drag polar* shown on the right, which of course is a key parameter in any form of gliding flight.

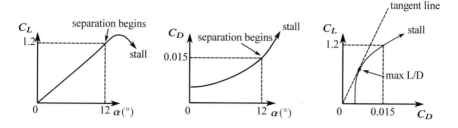

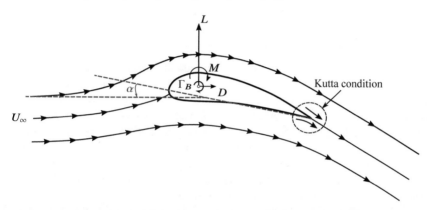

Fig. 5.7 Typical performance characteristics of a generic NACA0009 ($t/c = 0.09$) profile. For any given local geometry (profile), one can obtain characteristic data with which to then apply across a more complex body or shape through a strip-theory approach

Fig. 5.8 The Kutta–Joukowski theorem can be used to calculate the lift per unit span (in a strip-theory approach), where lift is shown to be proportional to the bound circulation Γ_B. The Kutta condition at the trailing edge dictates that fluid follows parallel to the sharp trailing edge

5.2 Kutta–Joukowski Theorem and the Generation of Lift

Let's begin by considering the forces and moments on an airfoil satisfying the Kutta condition, as shown in Fig. 5.8. Lift is a key ingredient, not only to flight when overcoming gravity, but as an efficient means to generate a side force for underwater maneuvering (think of a dolphin's fluke as a rudder). Although at this stage we are focusing on steady cases, we will later see how lift also contributes to thrust, i.e., have you stopped to consider what a penguin may be doing when *flying* underwater?

There are in fact many circular approaches to explain the generation of lift, ranging from those that are flat-out incorrect (Bernoulli) to those based on streamline curvature (Babinsky 2003). The classic pilot's approach, assuming that the top (or suction-side) surface is longer, and that a pair of fluid elements then *must* meet again at the trailing edge, provokes the use of Bernoulli (Eq. (5.5)) but can easily be discredited when considering the characteristic slender outer sections of bird wings. For the case of a very thin, yet highly cambered bird-wing profile, the upper and

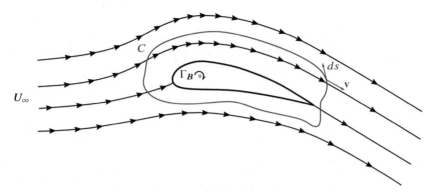

Fig. 5.9 Circulation can be calculated for any closed curve C, where ds is the local element of the curve with velocity **v**

lower surfaces have effectively the same length, and hence Bernoulli would state that no lift can be generated. However, we know this cannot be the case. Instead, let us take a more rigorous approach and explore the origins of what is known as the Kutta–Joukowski theorem (Kutta 1902; Joukovsky 1906). As shown in Fig. 5.9, we can relate local lift to the *bound vortex* or bound circulation Γ_B, which, as we will see later, is a fictitious yet convenient concept accounting for the net circulation of the boundary layers. In any event, we must assume that the Kutta condition holds at all times, i.e. that the trailing edge is sharp and that both top and bottom fluid layers meet parallel to one another at this singular point.

We first must recall the definitions of vorticity and circulation that were discussed in Chap. 2. To start, vorticity is defined as the curl of the velocity field:

$$\boldsymbol{\omega} = \nabla \times \mathbf{v} = \begin{vmatrix} \hat{i} & \hat{j} & \hat{k} \\ \frac{\partial}{\partial x} & \frac{\partial}{\partial y} & \frac{\partial}{\partial z} \\ u & v & w \end{vmatrix} \tag{5.4}$$

$$= \left(\frac{\partial w}{\partial y} - \frac{\partial v}{\partial z} \right) \hat{i} + \left(\frac{\partial u}{\partial z} - \frac{\partial w}{\partial x} \right) \hat{j} + \left(\frac{\partial v}{\partial x} - \frac{\partial u}{\partial y} \right) \hat{k}.$$

When a region of flow has $\boldsymbol{\omega} = 0$, then the flow field is referred to as *irrotational*. We can apply *potential-flow* theory to streamlines in the flow such that they have the same energy constant (recall Bernoulli's principle):

$$p + \frac{\rho U^2}{2} + \rho g z = const. \tag{5.5}$$

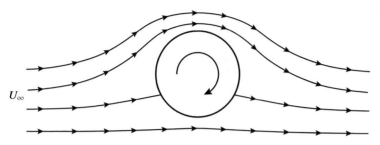

Fig. 5.10 Streamlines depicting superposition of freestream and point-vortex flow past a spinning cylinder (or sphere)

The circulation Γ, in turn, can be defined through the line integral of velocity around any closed curve in the flow:

$$\Gamma = - \oint_C \mathbf{v} d\boldsymbol{\ell}, \tag{5.6}$$

where $d\boldsymbol{\ell}$ is tangential to the arbitrarily chosen curve C, as depicted in Fig. 5.9, that encloses some region of vorticity. Circulation is a kinematic property dependent only on the velocity field and the choice of curve C.

Consider the flow past a spinning cylinder (or ball) in Fig. 5.10. Circulation around the body does not imply that fluid elements actually move around a loop per se. Rather we use linear superposition from potential flow to add (superimpose) both freestream and point-vortex contributions with one another. Using Stokes' theorem, we can relate circulation to vorticity:

$$\Gamma = - \oint_C \mathbf{v} d\boldsymbol{\ell} = - \iint_A (\nabla \times \mathbf{v}) d\mathbf{s}, \tag{5.7}$$

where of course $\boldsymbol{\omega} = \nabla \times \mathbf{v}$, and A represents the area enclosed by curve C.

Let us now consider an infinite *cascade* or row of airfoils turning the flow, as illustrated in Fig. 5.11. Symmetry requires that for faces **2** and **4**:

1. Pressure distributions are equal.
2. There is no net flux of x- or y-momentum.
3. Their contribution to Γ around the control volume will cancel.

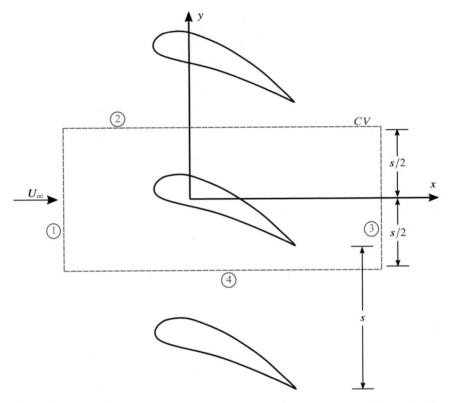

Fig. 5.11 A cascade of airfoils separated by vertical distance s responsible for turning the freestream flow downwards

We can therefore balance the y-momentum as follows (assuming the flow is steady, $\frac{d}{dt} \to 0$):

$$
\begin{aligned}
-F_y &= \frac{d}{dt}\int_{CV} \rho v \, d\mathcal{V} + \int_{CS} \rho v (\mathbf{v} \cdot \mathbf{n}) \, dA \\
&= \rho \int_{CS_1} (v_1)(-U_\infty - u_1) \, dA_1 + \rho \int_{CS_3} (v_3)(U_\infty + u_3) \, dA_3 \\
&= \rho \int_{-s/2}^{+s/2} (-v_1 U_\infty - v_1 u_1) \, dy + \rho \int_{-s/2}^{+s/2} (v_3 U_\infty + v_3 u_3) \, dy,
\end{aligned}
\tag{5.8}
$$

where the numbered subscripts represent the different edges of the CV according to Fig. 5.11. The balancing force in the y-direction is known as *lift* (L) and can be written in terms of the circulation by simplifying Eq. (5.8):

$$
L = \rho U_\infty \Gamma - \rho \int_{-s/2}^{+s/2} (u_3 v_3 - u_1 v_1) \, dy.
\tag{5.9}
$$

Similarly, we can perform an x-momentum balance, where now we must account of the pressure difference between faces **1** and **3**:

$$-F_x + \int_{CS} \rho d A = \frac{d}{dt} \int_{CV} \rho u d \Psi + \int_{CS} \rho u (\mathbf{v} \cdot \mathbf{n}) d A$$

$$-F_x = -\int_{CS_1} p_1 d A_1 + \int_{CS_3} p_3 d A_3$$

$$+ \rho \int_{CS_1} (U_\infty + u_1)(-U_\infty - u_1) d A_1$$

$$+ \rho \int_{CS_3} (U_\infty + u_3)(U_\infty + u_3) d A_3 \qquad (5.10)$$

$$= \int_{-s/2}^{+s/2} (p_3 - p_1) dy - \rho \int_{-s/2}^{+s/2} (U_\infty + u_1)^2 dy$$

$$+ \rho \int_{-s/2}^{+s/2} (U_\infty + u_3)^2 dy.$$

The resultant force in the x-direction is the *drag* (D), but what is $p_3 - p_1$? Using Bernoulli's equation (Eq. (5.5)), we can write each p in the form:

$$p = \frac{1}{2} \rho U_\infty^2 - \frac{1}{2} \rho \left[(U_\infty + u)^2 + v^2 \right]$$

$$= -\rho \left[U_\infty u + \frac{1}{2} u^2 + \frac{1}{2} v^2 \right], \qquad (5.11)$$

such that Eq. (5.10), where the force in the x-direction is the *drag D*, becomes

$$D = \frac{1}{2} \rho \int_{-s/2}^{+s/2} [u_1^2 - u_3^2 + 2U_\infty(u_1 - u_3) - v_1^2 + v_3^2] dy. \qquad (5.12)$$

Simplifying Eq. (5.9), we see that $u_3 v_3 - u_1 v_1$ tends to zero and that lift simply relates as

$$L = \rho U_\infty \Gamma. \qquad (5.13)$$

Furthermore, $u_1(y) = u_3(y)$ and $v_1(y) = -v_3(y)$, so the integrand in Eq. (5.12) tends to zero:

$$D = 0. \qquad (5.14)$$

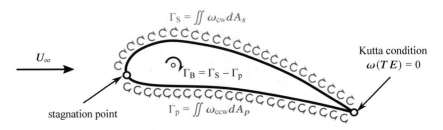

Fig. 5.12 The boundary layer on the top (suction) and bottom (pressure) sides of the airfoil can be modeled as clockwise (red) and counterclockwise (blue) vortex sheets, respectively. Note that the vorticity at the trailing edge (Γ_{TE}) should be zero in order to satisfy the Kutta condition

These two results do not change if the limits in y tend to $\pm\infty$ such that, for a body in isolation, we have the Kutta–Joukowski theorem ($L = \rho U_\infty \Gamma$) as depicted by the bound vortex (and the circulation Γ_B) in Fig. 5.8.

In theory, there are an infinite number of values for Γ at a given angle of attack α. However, thanks to viscosity, the Kutta condition (i.e., flow moves parallel at the sharp trailing edge) must be satisfied, in turn setting a specific value of Γ_B for a given set of in-flow conditions, e.g., α and U_∞. It is the requirement of the Kutta condition that helps fix the strength (vorticity distribution) of the boundary layers upstream, as depicted in Fig. 5.12. The bound circulation about the aerodynamic center of the airfoil is a function of the difference of the vorticity along either surface. Paradoxically, lift, as generated by such an airfoil, remains as an inviscid concept that can be predicted with potential-flow theory (recall the Euler equation from Chap. 2), which however simultaneously requires viscosity to enforce the Kutta condition at the trailing edge! In other words, without a sharp trailing edge, and viscosity limiting infinite acceleration around this edge, there would be no net lift.

5.2.1 Exercise on the Origins of Flight

Soaring birds, such as the albatross, are not the only ones looking to maximize their overall glide ratio (L/D). Flying squirrels, jumping snakes and even flying fish are continuously pushing the boundaries against gravity. In this current example, however, we will revisit the dawn of avian flight. Note that this process represented just one of three distinct pathways to have powered vertebrate flight in the skies, making for yet another compelling example of evolutionary convergence.

Although extant species like flying squirrels provide a compelling argument for the origins of flight in the forest canopy, one in which the animal climbs up trees and then subsequently glides down, in many therapod species like *Microraptor*, as shown in Fig. 5.13, it is difficult to envisage how they would manage to climb trees with their bulky—and unique!—pair of hind wings. Instead, it is hypothesized

Fig. 5.13 A microraptor's hind wings significantly increase its wing area (and hence BW), a feature that most likely allowed it to take to the air via Wing-Assisted Incline Running (WAIR). Note that in this problem we are effectively combining front and hind wing pairs into one giant lifting surface

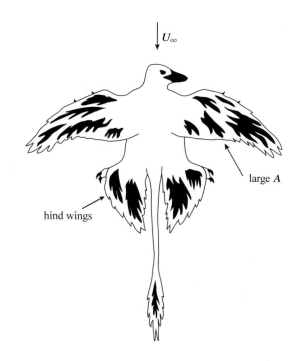

that these *paravian* dinosaurs may have developed powered flight through a transition from flap running or wing-assisted leaping, both enhancing ground-based performance by including aerodynamic forces from feathered wings, to what is termed Wing-Assisted Incline Running (WAIR); see Dececchi et al. (2016). In the case of WAIR, flapping wings can be used to reduce body weight through aerodynamic lift, where we define the ratio of aerodynamic lift to gravitational pull as

$$BW = \frac{C_L \rho U_\infty{}^2 A}{2mg}. \tag{5.15}$$

C_L, A, and m are the total planform lift coefficient, wing area, and mass, respectively; U_∞ is the running velocity (before take-off); and g, of course, is the gravitational acceleration. Naturally with $BW > 1$, we can envision *Microraptor* taking to the air, and hence bypassing the need (or challenge) to climb up into the forest canopy.

It turns out that *Microraptor*, unlike many other *paravian* dinosaurs, with its unique set of two wing pairs, had a relatively low wing loading, i.e., a large wing area relative to its body mass, thanks in part to its feathered legs. If we take one specimen with tip-to-tip wing span of 0.5 m, estimate values such as $A = 0.4\,\text{m}^2$ and $m = 0.2\,\text{kg}$, and then assume a ground-based running speed from that of an equivalently sized, extant turkey at $U_\infty \approx 2.5\,\text{m/s}$, we can back out the necessary body lift coefficient to become airborne. Our back-of-the-envelope calculation gives us a value of $C_L \approx 1$, which is a completely reasonable value for such a wing planform (note that we are effectively combining front and hind wing pairs into one giant lifting surface). In other words, it seems reasonable that *Microraptor* was able to use WAIR to run and take-off like extant birds today, potentially indicating a pathway taken in that evolutionary transition.

5.3   Unsteady (Planar) Wakes

In practice, most biological systems employ unsteady motion (e.g., flapping or oscillatory movement) as their *modus operandi*. The acceleration term in the Navier–Stokes equation (Eq. (2.47)) becomes the rule rather than the exception. As per the discussion on strip theory above, we can envision how the variation in time-dependent lift, thrust, etc. would vary as a function of the section position on, say, a wing, where the root region would see a smaller excursion than the tip. Nevertheless, we will again keep it simple and focus on one particular section undergoing a local simplified motion.

Let us begin by characterizing the influence of unsteadiness on the wake of this local section experiencing pitching ($\dot{\alpha}$) and plunging/heaving ($\dot{h}$) motions, as shown in Fig. 5.14. If we start by defining a simple unsteady variation in the profile plunge position $h(t)$, defined by

$$h(t) = h_0 \cos(2\pi f t), \tag{5.16}$$

where h_0 represents the amplitude of motion (typically the trailing-edge excursion), we can describe a broad range of propulsive cases. Unlike the steady world, the wake structure behind these oscillating foils is primarily described by the Strouhal number (St), rather than say the Reynolds number, which can be parametrized as

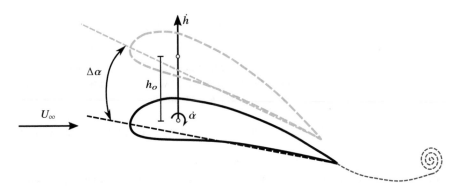

Fig. 5.14 Profile experiencing pitching ($\dot{\alpha}$) and plunging/heaving ($\dot{h}$) motions. For the case of a plunging motion the trailing-edge excursion is defined as $2h_0$

follows:

$$St = \frac{f(2h_o)}{U_\infty}, \tag{5.17}$$

where $2h_o$ represents the trailing-edge excursion and, hence, the global width of the propulsive wake. Further to the Strouhal number, Birnbaum (1924a, 1924b) proposed an additional dimensionless group known as a *reduced frequency* (k), which is a measure of the relative degree of unsteadiness in a flow. The reduced frequency, unlike the Strouhal number, allows us to compare the flow's characteristic convective time (Δt_{conv}) to the characteristic time associated with the plunge oscillation itself (Δt_{osci}):

$$k = \frac{\Delta t_{conv}}{\Delta t_{osci}} = \frac{(\frac{c}{2})/U_\infty}{1/2\pi f} = \frac{\pi f c}{U_\infty}, \tag{5.18}$$

where, as before, c represents the profile chord length. Typically we define $k \leqslant 0.05$ as quasi-steady, where flow acceleration can be considered negligible. In contrast, when $k \geqslant 0.05$, the problem begins to show a dependence on the time-history of the flow. Of course, this threshold is somewhat arbitrary and a function of geometry, etc.

Lord Kelvin (Thomson 1868) considered the case of an inviscid, barotropic flow with conservative body forces:

$$\mathbf{f} = -\nabla F, \tag{5.19}$$

where F here represents the body-force potential. Kelvin's circulation theorem shows that the circulation around a closed fluid line remains constant with respect to time (i.e., $\Gamma_2 = \Gamma_1 = 0$), as shown in Fig. 5.15.

When we consider the material derivative of circulation within a closed path, we obtain

$$-\frac{D\Gamma}{Dt} = \frac{d}{dt}\left(\oint_c \mathbf{v} \cdot d\mathbf{s}\right) = \oint_c \frac{d\mathbf{v}}{dt} \cdot \mathbf{n}d\mathbf{s} + \oint_c \mathbf{v} \cdot \frac{d}{dt}(d\mathbf{s}), \tag{5.20}$$

where

$$\frac{d}{dt}(d\mathbf{s}) = d\mathbf{v}. \tag{5.21}$$

Furthermore, for an inviscid flow:

$$\frac{d\mathbf{v}}{dt} = \mathbf{f} - \frac{1}{\rho}\nabla p = -\nabla F - \frac{1}{\rho}\nabla p, \tag{5.22}$$

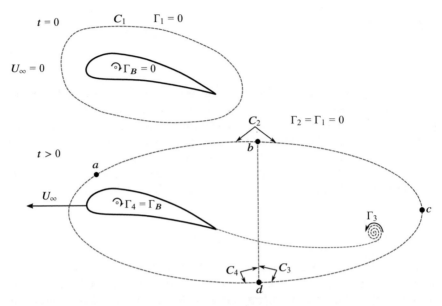

Fig. 5.15 Kelvin's circulation theorem showing net circulation at two distinct time steps, where Γ_1 is at start of motion from rest and Γ_2 represents a later time once a wake vortex is formed. Initially the fluid is at rest relative to profile. Once moving, the net change on total contour *abcda* is zero ($\Gamma_2 = \Gamma_1 = 0$). It can also be ascertained that $\Gamma_4(abda)$ and $\Gamma_3(bcdb)$ are opposite: $\Gamma_3 = -\Gamma_4$

such that

$$\frac{d}{dt}\left(\oint_c \mathbf{v} \cdot d\mathbf{s}\right) = -\oint_c dF - \oint_c \frac{dp}{\rho} + \oint_c \mathbf{v} \cdot d\mathbf{v}. \qquad (5.23)$$

All three integrals on the right-hand side of Eq. (5.23) tend to zero, such that we obtain

$$\frac{d}{dt}\left(\oint_c \mathbf{v} \cdot d\mathbf{s}\right) = 0, \qquad (5.24)$$

or simply

$$\frac{D\Gamma}{Dt} = 0. \qquad (5.25)$$

Equation (5.25) represents what is known as Kelvin's circulation theorem (Thomson 1868). This is an important result, as we will see later, the circulation generated on the profile responsible for lift (and thrust) will leave an equal and opposite contribution of circulation in its wake.

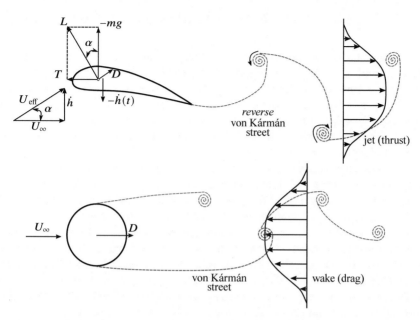

Fig. 5.16 Kelvin's theorem provides insight into the generation of a *reverse* von Kármán street for a plunging profile, shown in contrast here to the wake behind a cylinder. Furthermore, the *Knoller–Betz effect* provides insight into the efficient propulsion of a broad variety of animals through what is known as lift-based propulsion. Of particular interest is the leading-edge suction responsible for producing thrust (T), like in Sect. 5.1, but now varying as a function of the time-dependent motion

Let us apply Kelvin's circulation theorem to a simple example of a plunging airfoil, which is representative of a wing section in flapping flight or a fin section in swimming. Here, we can observe the distinct roll up of boundary-layer vorticity from both the top (suction) and bottom (pressure) sides of the airfoil into a *reverse* von Kármán street (see Fig. 5.16). We also observe a simple yet clear description of how the instantaneous tilting of resulting lift L contributes to a propulsive force (i.e., thrust T), where the tilting of the lift vector is based on the time-varying change in the oncoming flow (U_{eff}) rather than just the relative ground speed of the body (U_∞) in this unsteady case. This generalized description of unsteady propulsion, known as the *Knoller–Betz effect* (Knoller 1909; Betz 1912; Katzmayr 1922), provides a clear basis with which animals propel themselves efficiently through various fluid mediums. This leading-edge suction effect, responsible for overall system thrust, as per Sect. 5.1, is now varying in time as a function of the time-dependent motion.

5.3.1 Exercise on Simultaneous Generation of Lift and Thrust

The common argument against using a bio-inspired approach to propulsion is that Nature—with a few known exceptions that we will touch on later—must oscillate a heavy appendage in order to generate thrust, as shown above in Fig. 5.16. This oscillatory motion, whether on an appendage, or even the continuous deformation of a jellyfish bell, will require a high expenditure of energy to account for inertial forces, at least when compared to the elegance of a continuously rotating system like that of a propeller.

However, there is one key misconception in this argument, and it is based on the decomposition of lift and thrust (for a heavier-than-air flying system), as shown in Fig. 5.17. Let us return to the *paravian* dinosaurs and their first experiments with powered flight. Their challenge, of course, was to overcome gravity, but once airborne in order to maintain lift (without gliding down from the forest canopy), they also had to generate thrust so as to overcome drag. In conventional aircraft, the lifting surface(s) (i.e., the wings) are separate to the propulsion system, be it the propellers, jet engines, etc. Since each individual system (e.g., blade, wing, control surface, etc.) has its respective boundary layers (source of frictional drag), all this vorticity, once shed into the wake, is forever lost and therefore wasted.

In contrast, biological systems *recycle* boundary-layer vorticity by shedding this rotational energy into distinct wake structures during each wing stroke. Just look at the formidable aerial acrobatics of dragonflies, in which vortices shed from their front pair of wings can be repurposed on the hind pair (and *vice versa*), as shown in Fig. 5.18. These shed vortex structures serve many purposes, and for instance, when impinging on downstream surfaces such as *Microraptor's* hind wings, these wake vortices would likely provide an additional opportunity to enhance the system's overall efficiency. Leave it to Nature to use energy in the most sparing yet clever manner, a key lesson we might want to consider when marveling at our seemingly clunky engineered analogs.

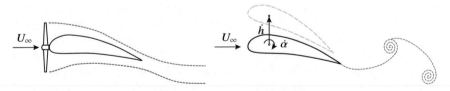

Fig. 5.17 (left) Cartoon depicting the decomposition of lift and thrust in heavier-than-air flying systems where all wake vorticity from lifting and propulsive surfaces is shed in an incoherent manner. (right) Here a flapping wing (section) rolls up boundary-layer vorticity into coherent structures (wake vortices) that can be exploited downstream by other appendages or neighboring animals (e.g., fish schooling)

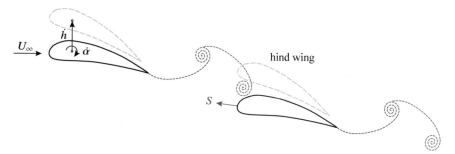

Fig. 5.18 Tandem-flapping wing (appendage) arrangement is yet another example of evolutionary convergence, i.e. this arrangement is observed from dragonflies all the way to *Plesiosaurs*, providing a means to recycle shed vortices from the leading wing and generate additional thrust through leading-edge suction (*S*) on the hind wing. Note that these types of interactions also describe interactions in groups such as fish schooling and that the wake vortices can also be recycled to enhance lift/side force, etc.

5.4   Decomposition of Forces (Circulatory vs. Added Mass)

So, how do we estimate the instantaneous forces on, for example, the moving profile described in the previous sections? If we return to control-volume analysis and consider the vertical, *y*-component of momentum:

$$F_y = -L = \frac{d}{dt} \iiint_{CV} \rho v d\mathcal{V} + \iint_{CS} \rho v (\mathbf{v} \cdot \mathbf{n}) dA, \tag{5.26}$$

we can no longer neglect the unsteady term. In fact, the acceleration field is its highest next to the accelerating profile surface. In order to create a simple, tractable model for these unsteady aero/hydrodynamic problems, we follow the steps traditionally taken in potential-flow theory, and break the instantaneous force into two distinct contributions:

1. Circulatory forces and
2. Non-circulatory (added-mass) forces

where the latter is sometimes simply defined as all force *in phase* with the body acceleration.

5.4.1   Circulatory Force

Starting with the circulatory forces, von Kármán and Sears (1938) expanded on the Kutta–Joukowski theorem and showed how best to relate the rate of change of vertical momentum imparted by a vortex pair (see the bound and wake (starting) vortices in Fig. 5.19, where the two vortices are represented by the subscripts *B* and

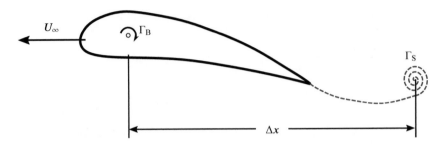

Fig. 5.19 The kinematics of the time-varying bound vortex with circulation Γ_B, and the growing wake vortex with circulation Γ_S, can be related to the instantaneous lift L_{circ}; see von Kármán and Sears (1938)

S, respectively), to the magnitude of circulatory lift:

$$
\begin{aligned}
L_{circ} &= \frac{d}{dt}[\rho \times \Gamma] \\
&= \rho[(u_B - u_S)\Gamma + (x_B - x_S)\dot{\Gamma}] \\
&= \rho U_\infty \Gamma + \rho \Delta x \dot{\Gamma},
\end{aligned}
\tag{5.27}
$$

where the circulation of the bound (Γ_B) and wake (Γ_S) vortices are equal: $\Gamma_B = \Gamma_S = \Gamma$ per Kelvin's theorem (Eq. (5.25)). In a similar manner, we can write the instantaneous circulatory lift ($C_{L_{circ}}$) in dimensionless form:

$$
C_{L_{circ}} = \frac{2}{U_\infty^2 c}[U_\infty \Gamma + \Delta x \dot{\Gamma}].
\tag{5.28}
$$

To develop a general appreciation for how long a starting vortex with circulation Γ_S, when left in the wake, affects the overall sectional lift, we can look to Wagner (1925), who elegantly showed how lift asymptotically reached steady-state conditions, after starting from rest, once at a distance $\Delta x \approx 10c$ downstream, as shown in Fig. 5.20. In other words, a continuously pitching and plunging profile will never see *steady* conditions as per the simplified representations of Sect. 5.2.

5.4.2   Added-Mass Force

Let us now consider the non-circulatory (or added-mass) force, which is generally described as a force in phase with acceleration. If a body (mass m), when immersed in a fluid medium with density ρ, is accelerated, it will displace and accelerate some volume of surrounding fluid with it. The total reaction force associated with

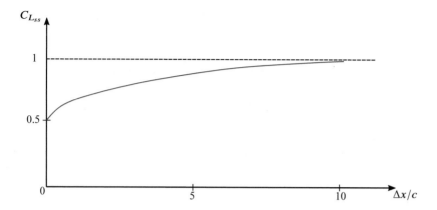

Fig. 5.20 Wagner (1925) showed that a starting vortex with circulation Γ_S affects overall sectional lift for as much as $\Delta x \approx 10c$ into the wake

accelerating the body *and* the affected fluid is simply:

$$F = (m + \rho V)a, \tag{5.29}$$

where V is the affected fluid volume. The added-mass force (F_{AM}) is defined as the part of the above reaction force required to increase the kinetic energy of the surrounding fluid, where:

$$F_{AM} = (\rho V)a. \tag{5.30}$$

Generally, the kinetic energy (KE) associated with the motion of the surrounding fluid is

$$KE = \frac{\rho}{2} \int_{\infty} (u^2 + v^2 + w^2)dV, \tag{5.31}$$

where V is the entire volume of fluid affected by the body motion. If the motion is a simple, steady translation (at velocity U), then KE becomes

$$KE = \rho \frac{I}{2} U^2 \quad , \text{ where } \quad I = \int_{\infty} \left(\frac{u^2 + v^2 + w^2}{U^2} \right) dV. \tag{5.32}$$

Note that the integral I is constant in a potential-flow solution. If the body were now to be accelerated, the fluid's kinetic energy will (most likely) increase as the body velocity changes. Work must be done on the fluid to provide this increase in energy, which is proportional to the rate of change of the kinetic energy:

$$\frac{dKE}{dt} = -F_{AM}U, \tag{5.33}$$

where F is effectively an additional drag acting on the body and (from Eqs. (5.32) and (5.33)):

$$F_{AM} = -\rho I \frac{dU}{dt}.$$ (5.34)

We should quickly recognize this as the form $F = ma$, where $m = m_a = \rho I$ represents the mass of fluid, or *added mass*, being displaced/accelerated by the motion of the body.

For instance, we can show using potential-flow theory that the added-mass force associated with the acceleration of a circular plate, perhaps the simplest approximation to a simple appendage required for drag-based propulsion or maneuvering, can be defined as

$$m_a = \frac{8}{3} \rho r^3,$$ (5.35)

where r is the radius of the plate and the added-mass force is simply the product of m_a and the acceleration of the plate: $F_{AM} = m_a a$. Similarly, the added mass associated with the acceleration of a spherical body, say for instance an escaping octopus, is known to be

$$m_a = \frac{2}{3} \pi \rho r^3.$$ (5.36)

5.4.3 Exercise on Gust Response of Milkweed Seed

As a simple illustration of added mass in Nature, consider a spherical milkweed seed released (or launched) by a gust of air, as shown in Fig. 5.21. In reality, the bristled seeds form a porous body, but at sufficiently low Reynolds number, act to effectively create a lightweight, non-porous sphere. The force required to launch the seed from rest will be a function of the added-mass force, noting that the wake vortices produced behind an accelerating spherical body are generally weak, and as such, the circulatory forces can be ignored here.

If we imagine a milkweed seed that has just been released from its pod (i.e., initially at rest) via a vertical gust of air, let us calculate how quickly it takes for the seed to be launched. Let us assume that the seed has a diameter $d = 5\,\text{cm}$ and a mass of only $m = 3\,\text{mg}$ and that the gust can be defined by a linear ramp $V(t) = a_f t$, where $a_f = 0.5\,\text{m/s}^2$ is the acceleration of the fluid. We can then draw a simple force balance, as shown on the left in Fig. 5.21. The net force on the body

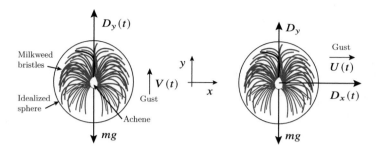

Fig. 5.21 At sufficiently low Reynolds numbers the milkweed seed's achene and bristles can be approximated by solid (non-porous) sphere. On the left, the seed is lofted into the air by a vertical gust. Then once airborne and falling at terminal (steady-state) velocity, the seed is then hit by second horizontal gust, as shown on the right

is a balance between its drag and weight:

$$F_{net} = ma_b = D_y(t) - mg$$

$$ma_b = m_a(a_f - a_b) + \frac{1}{2}\rho C_D(V - v_b)^2 A - mg \qquad (5.37)$$

$$= m_a a_f - m_a a_b + \frac{1}{2}\rho C_D(V - v_b)^2 A - mg,$$

where the drag force $D_y(t)$ is the sum of the added-mass force and the quasi-steady drag approximation, where $C_D = 0.5$ represents the steady-state drag coefficient for a sphere with cross-sectional area A. Rearranging, we come to an expression for the instantaneous acceleration of the seed a_b:

$$a_b(t) = \frac{m_a a_f + \frac{1}{2}\rho C_D(V - v_b)^2 A - mg}{(m + m_a)}, \qquad (5.38)$$

where m_a is the added mass as defined in Eq. (5.36) for a sphere. Assuming an initial body velocity (v_b) and acceleration of zero, we can iteratively solve for a_b, v_b, and $D_y(t)$ over time, as plotted in Fig. 5.22. Initially, as the gust begins to ramp up, the seed falls under its own weight, as the gust is incapable of accelerating both the seed and its surrounding fluid upwards instantaneously. Shortly thereafter, however, the milkweed seed begins to accelerate in the (upward) direction of the gust.

Assuming now that, after being lofted into the air by a vertical gust, the milkweed seed begins its descent back toward the earth at its terminal (steady-state) velocity. At a height of 44 cm the seed is fortuitously hit by a horizontal gust ($U(t)$). We can now compare the seed's trajectory to the ground comparing both the added mass as well as a simpler, quasi-steady drag model (the latter does not account for the true unsteady nature of the flow). Figure 5.21 shows this new arrangement where now the drag force is horizontal (D_x), in the direction of the gust, and the seed's weight is balanced by steady air resistance (i.e., there is no longer a vertical acceleration).

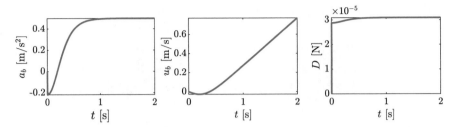

Fig. 5.22 Through an iterative calculation we can plot the milkweed seed's acceleration, as well as its velocity and drag, over time

Assuming the gust follows the same linear ramp as before but now in the horizontal direction where $U = a_f t$, the x- and y-components of the seed's acceleration, using added-mass (AM) and quasi-steady (QS) models, can be written:

$$a_{AM,x}(t) = \frac{m_a a_f + \frac{1}{2}\rho C_D (U - u_b)^2 A}{(m + m_a)},$$

$$a_{QS,x}(t) = \frac{\frac{1}{2}\rho C_D (U - u_b)^2 A}{m}, \tag{5.39}$$

$$a_{b,y} = \frac{\frac{1}{2}\rho C_D v_t^2 A - mg}{m} = 0,$$

where v_t is the terminal velocity of the seed:

$$v_t = -\left(\frac{2mg}{\rho C_D A}\right)^2 = const. \tag{5.40}$$

As for the initial launch, the time-resolved acceleration and velocity of the seed can now be plotted using each model, see Fig. 5.23. Additionally, we can plot the trajectory of the seed as it is blown laterally during its descent using

$$x_b(t) = x_b(t-1) + u_b(t)\Delta t + \frac{1}{2}a_x(t)\Delta t^2, \quad \text{and}$$

$$y_b(t) = y_b(t-1) + v_t \Delta t, \tag{5.41}$$

where x_b and y_b represent the horizontal and vertical positions of the seed at some time t after being hit by the horizontal gust.

In Fig. 5.23, we see a stark difference between the acceleration using each of the added-mass and quasi-steady models; the quasi-steady approach depends only on the instantaneous velocity of the body and fails to capture the sudden acceleration that the seed in fact experiences. As a result, when added mass is accounted for, the

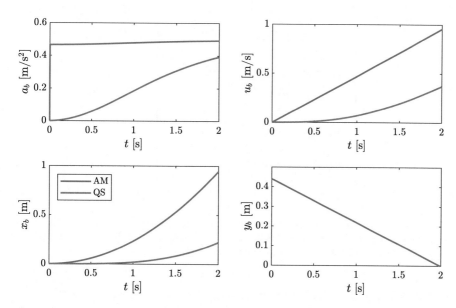

Fig. 5.23 The acceleration, velocity, and trajectory of a milkweed seed hit by a horizontal gust are predicted and compared using added-mass (blue) and quasi-steady (red) models

predicted horizontal travel of the seed is drastically increased, which is, of course, important for the dispersion and proliferation of the species in question.

5.5 Rapid Area Change

For many dynamics problems, we often tend to assume that the added mass of fluid around an immersed body remains constant over the acceleration. However, in practice, the amount of fluid affected by the body acceleration can very much change over time, particularly in systems with time-varying shapes. In these cases, the one-dimensional added-mass force should account for a changing added mass and be written as the time derivative of the product of the mass and its velocity:

$$F_{AM} = \frac{d}{dt}(m_a v) = m_a a + \dot{m}_a v, \tag{5.42}$$

where $\dot{m}_a$ is the rate of change of added mass.

The amount of added mass can change due to a number of factors including a change in body shape, such as the rapid change in frontal area of a *pitching* airfoil (basic approximation to perching). This pitching motion produces a rapid change in the frontal area as illustrated by the perching maneuver of a large bird of prey, as

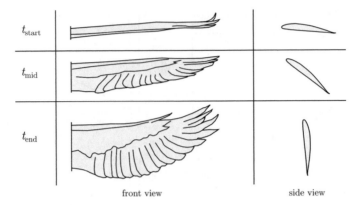

shown in Fig. 5.24, which can be characterized by a *shape change number*:

$$\Xi = \frac{V^2}{aL},$$ (5.43)

where V is the velocity associated with the area change, a the body deceleration, and L the characteristic length scale (wing chord in this case). Weymouth and Triantafyllou (2013) showed that varying the shape change number can drastically alter the added mass associated with the acceleration (or deceleration), and therefore the global forces acting on the body. This shape change feature is pervasive across biological propulsion and has yet to receive sufficient consideration—hence our current understanding is incomplete.

5.5.1 Exercise on Perching

Now let us consider the perching maneuvers observed in much smaller (and lighter) birds such as a chickadee rapidly opening up its wings from an initial ballistic phase, as shown here in Fig. 5.25. Recall the squared-cubed scaling for birds discussed (and shown) in the first chapter. By increasing the frontal area exposed to the incoming flow as quickly as possible, all the while decelerating, the chickadee is able to rapidly transfer kinetic energy into the wake (fluid) and slow down within a few body lengths. The motion of the wings during the perching maneuver can for now be approximated as an airfoil undergoing a steady deceleration along with a pitch-up rotation (in reality the wings fan out laterally from the ballistic body shape and the flow is three-dimensional). Unlike the milkweed seed in Sect. 5.4.3, the asymmetry

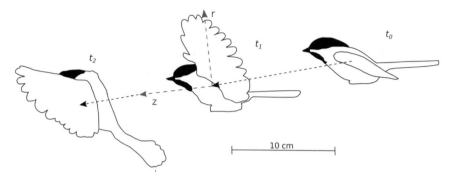

Fig. 5.25 Rapid opening of wings on a chickadee coming into perch. Initially the chickadee is in ballistic phase with wings tucked (t_0), but soon after deploys its wings (t_1), in turn generating enormous added-mass and circulatory forces when coming into perch (t_2). Figure reproduced from Polet and Rival (2015)

of the wing and the rapid shape change implies that the forces acting on the body will be a function of added mass but also of circulatory effects.

We can prescribe the steady deceleration (from U_0 to 0) and 90° rotation (from angle of attack $\alpha = 0$ to $\pi/2$) using the following:

$$u(t) = U_0(1 - t^*), \quad \text{and} \tag{5.44}$$

$$\alpha(t) = \frac{\pi}{2}\left[t^* - \frac{\sin 2\pi t^*}{2\pi}\right], \tag{5.45}$$

where t^* is the dimensionless time $t^* = t/T$ normalized by the maneuver period T. The wing velocity and rotation are plotted over the period in Fig. 5.26.

As described in Sect. 5.4, the aerodynamic forces on the body can be decomposed into an added-mass force, in this case associated with a rapid change in shape, and a circulatory force, which arises from the production of strong leading-edge (bound) and wake vortices. Let us begin by considering the added-mass force.

By approximating the chickadee wing as a flat plate, we can apply the (potential) function derived by Milne-Thomson (1968) for an accelerating and rotating flat plate to find the added-mass force coefficient:

$$C_{AM} = \frac{\pi c}{2U_0^2}\left[\dot{\alpha} u \cos\alpha + \dot{u}\sin\alpha + c\ddot{\alpha}(1/2 - x_p)\right], \tag{5.46}$$

where $\dot{\alpha}$ and $\ddot{\alpha}$ are the first and second temporal derivatives of the angle of attack describing the pitch-up motion, respectively, and $x_p = 1/6$ is the distance from the leading edge to the position of wing rotation, as a portion of the total chord c.

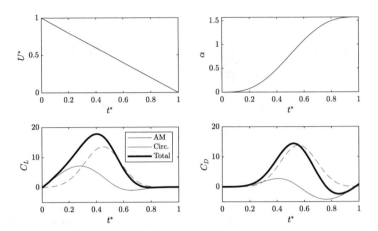

Fig. 5.26 Top row shows the overall perching motion. The bottom row breaks down the instantaneous lift and drag during the maneuver acting on the chickadee's wings for a shape change number of $\Xi = 0.5$. Figure reproduced from Polet and Rival (2015)

The total added-mass force, acting normal to the plate surface, can itself be divided into its lift and drag components (see blue curves in Fig. 5.26 for $\Xi = 0.5$):

$$C_{L,AM} = C_{AM} \cos \alpha;$$
$$C_{D,AM} = C_{AM} \sin \alpha. \tag{5.47}$$

Now, we will consider the circulatory forces generated by the pair of leading-edge (bound) and wake vortices attached to our flat-plate approximation; see Fig. 5.27. Recalling the definition of circulatory lift from von Kármán and Sears (1938) in Eq. (5.28), here, we will focus solely on the generation of circulation from the vortex dipole as it is the dominant term. The resultant circulatory force is therefore simply:

$$F_{\text{circ}} = -\rho c \dot{\Gamma}, \tag{5.48}$$

where c represents the spacing between growing vortices and $\dot{\Gamma}$ is the rate at which circulation is shed from the trailing edge forming a wake vortex.

While at an angle to the incoming flow, the velocity on the pressure side of the wing increases relative to that on the suction side. The velocity differential ($U_\perp$) at the trailing edge of the airfoil generates vorticity that rolls up into the trailing-edge vortex (in an attempt to satisfy the Kutta condition). This wake vorticity, and therefore circulation, increases at a rate equivalent to the velocity of the trailing edge (U_{TE}). The rate of change of circulation at the leading edge is equal and opposite as dictated by Kelvin's theorem (Eq. (5.25)), and the $\dot{\Gamma}$ term can be written in terms of

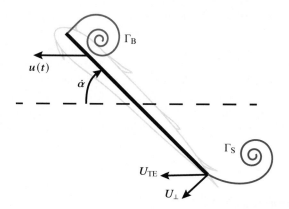

Fig. 5.27 The generation of circulation of equal magnitude but opposite direction, $\Gamma_B = \Gamma_S$, at the leading and trailing edges (of the flat-plate approximation) contributes to lift and drag forces during the chickadee's perching maneuver. Note that in maneuvering systems often leading-edge vortices (LEVs) are formed, and if stable, act as analogs to the traditional bound vortices discussed above for attached flow

$U_\perp$ and U_{TE}:

$$\dot{\Gamma} = U_{TE} U_\perp. \tag{5.49}$$

Both of these above velocities can, in turn, be written as functions of the magnitude and rate of change of the plate angle of attack:

$$U_{TE} = \sqrt{u^2 + 2u\eta_p c\dot{\alpha} \sin\alpha + \eta_p^2 c^2 \dot{\alpha}^2}, \quad \text{and} \tag{5.50}$$

$$U_\perp = u \sin\alpha + \eta_p c\dot{\alpha}, \tag{5.51}$$

where $\eta_p \equiv 1 - x_p$ and $x_p = 1.6$ for this exercise.

As with the added-mass force, the circulatory force in Eq. (5.48) results in a force coefficient,

$$C_{\text{circ}} = \frac{2U_{TE} U_\perp}{U_0^2}, \tag{5.52}$$

that can be decomposed into lift and drag coefficients, as was done in Fig. 5.26. Therefore, when adding both added-mass and circulatory components we obtain the following expression for the instantaneous resulting force:

$$
\begin{aligned}
C_F =& \frac{2}{U_0^2} \left[U \sin\alpha + c\dot{\alpha}(1 - x_p) \right] \sqrt{u^2 + 2uc\dot{\alpha}(1 - x_p)\sin\alpha + c^2\dot{\alpha}^2(1 - x_p)^2} \\
&+ \frac{\pi c}{2U_0^2} \left[U\dot{\alpha}\cos\alpha + \dot{u}\sin\alpha + c\ddot{\alpha}(1/2 - x_p) \right].
\end{aligned}
\tag{5.53}
$$

The total force coefficient for the perching maneuver is simply the superposition of the added-mass (Eq. (5.46)) and circulatory (Eq. (5.52)) forces. Figure 5.26 shows the total force as a function of time for each of lift and drag. Of particular note are the contributions of added mass. Early in the prescribed motion, the added mass generates large amounts of lift that dominates the early stages of the perching maneuver, keeping the chickadee airborne as it slows. However, later in the perch maneuver, the kinetic energy gained by the surrounding fluid mass is counter-productive and actually produces additional thrust that must be balanced by the circulation-based drag.

5.6   Root and Tip Vortices

Of course, in practice, no wing or appendage is infinitely long. Our previous example for the chickadee assumes a two-dimensional (strip-theory) approach, but in reality, these results will deviate as three-dimensional effects, such as root and tip vortices, begin to dominate. To address these effects for steady (static) conditions, Prandtl (1918) developed the *lifting-line theorem*. As we will see, this elegant approach allows one to distribute the local lift in a sectional (strip-theory) approach, allowing for the analogous accounting of bound circulation and the roll up of root and tip vortices in the wake.

Let us begin by considering a simpler case of a *gliding* ray, along with degrees of abstraction capturing the same spanwise lift distribution (Fig. 5.28). For the case of our steady (gliding) ray with elliptical lift distribution, we obtain

$$\Gamma(z) = -\Gamma_o \sqrt{1 - \left(\frac{z}{b/2}\right)^2},\tag{5.54}$$

where $\Gamma(z)$ is the distribution of circulation as a function of the spanwise distance (z) from the wing's centerline, Γ_o is the (peak) circulation at the centerline, and

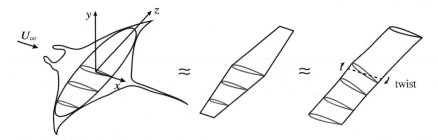

Fig. 5.28 A basic abstraction to a ray gliding in the ocean. In turn we can simplify further and use a tapered planform, as shown in the middle. Finally, we can also twist a constant-chord planform to produce an analogous lift distribution, as shown on right

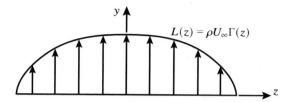

$$L(z) = \rho U_\infty \Gamma(z)$$

Fig. 5.29 An elliptical lift distribution produces this simple spanwise variation that is directly proportional to the local circulation, as per the Kutta–Joukowski theorem: $L(z) = \rho U_\infty \Gamma(z)$

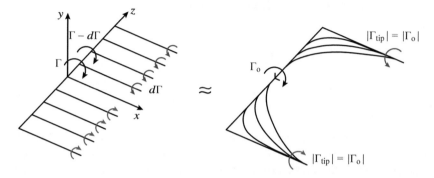

Fig. 5.30 The release of line (wake) vortices from the wing with elliptical lift distribution (left) has a constant strength ($d\Gamma$) across the span. On the right we see that these wake vortices roll up into distinct tip vortices such that $|\Gamma_{\text{tip}}| = |\Gamma_o|$

b is the span length. Recalling our definition of lift as a function of local (bound) circulation from the Kutta–Joukowski theorem (Eq. (5.9)), the lift produced by the *wing* follows the same shape as the variation in bound circulation $\Gamma(z)$; see Fig. 5.29. The spanwise variation in the circulation of shed vortices into the wake (vortex lines parallel to the freestream flow) has a strength equal to the variation in circulation on the wing itself. For our elliptical lift distribution, the shed vortex strength is equal ($d\Gamma$) across the span, as depicted in Fig. 5.30. In practice, these wake vortex lines roll up into coherent tip (or root) vortices such that $|\Gamma_{\text{tip}}| = |\Gamma_o|$.

Let us now calculate the induced velocity $v(z)$, referred to as *downwash*, for an arbitrary position z' along the span:

$$d\Gamma = \frac{d\Gamma}{dz'} dz'$$
$$= -\Gamma_o \frac{1}{2\sqrt{1 - (\frac{z'^2}{b/2})^2}} \left(\frac{-2z'}{b/2} \right) \frac{dz'}{b/2}. \tag{5.55}$$

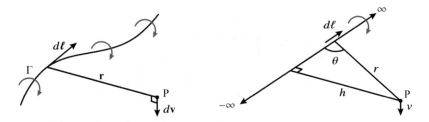

Fig. 5.31 The Biot–Savart law can be applied to an arbitrary vortex filament (line) following $d\boldsymbol{\ell}$ to determine the induced velocity $d\mathbf{v}$ at some position P away from the filament

Substituting in $\zeta' = \frac{z'}{b/2}$, we can simplify the above to

$$d\Gamma = \Gamma_o \frac{\zeta' d\zeta'}{\sqrt{1 - \zeta'^2}}. \tag{5.56}$$

In order to continue, let us return to the concepts of point vortices in Sect. 2.5. Analogous to electromagnetic induction, the *Biot–Savart law* (Biot & Savart 1820) describes the induced velocity field around a given vortex filament:

$$d\mathbf{v} = \frac{\Gamma}{4\pi} \frac{d\boldsymbol{\ell} \times \mathbf{r}}{|\mathbf{r}|^3}, \tag{5.57}$$

where $d\boldsymbol{\ell}$ is a vector along the vortex filament and $\mathbf{r}$ is the distance to some point P where a velocity $d\mathbf{v}$ is induced; see Fig. 5.31.

For a straight vortex filament of infinite length, we can calculate the induced velocity field by integrating Eq. (5.57) over all space:

$$\begin{aligned}
\mathbf{v} &= \int_{-\infty}^{\infty} \frac{\Gamma}{4\pi} \frac{d\boldsymbol{\ell} \times \mathbf{r}}{|\mathbf{r}|^3} \\
&= \frac{\Gamma}{4\pi} \int_{-\infty}^{\infty} \frac{\sin\theta}{r^2} d\ell,
\end{aligned} \tag{5.58}$$

where $d\boldsymbol{\ell} \times \mathbf{r} = d\ell \sin\theta$. Using simple trigonometric relationships (shown in Fig. 5.31), we can relate r and $d\ell$ to the angle θ:

$$r = \frac{h}{\sin\theta} \quad \text{and} \quad \ell = \frac{h}{\tan\theta} \quad \rightarrow \quad d\ell = -\frac{h}{\sin^2\theta} d\theta. \tag{5.59}$$

Substituting these relationships into Eq. (5.58), we find

$$
\begin{aligned}
v &= \frac{\Gamma}{4\pi h} \int_{-\infty}^{\infty} \frac{\sin\theta}{r^2}\, d\ell \\[4pt]
&= \frac{\Gamma}{4\pi h} \int_{0}^{\pi} \sin\theta\, d\ell \\[4pt]
&= \frac{\Gamma}{2\pi h}.
\end{aligned}
\tag{5.60}
$$

And hence, for a semi-infinite vortex filament (i.e., a tip-vortex),

$$
v = \frac{\Gamma}{4\pi h}.
\tag{5.61}
$$

Now, returning to our problem, where we would like to calculate the downwash ($v(z)$) some distance z' along the span of the wing, we can substitute Eq. (5.56) into Eq. (5.61):

$$
\begin{aligned}
dv &= \frac{d\Gamma}{4\pi h} = \frac{d\Gamma}{4\pi(z'-z)} \\[4pt]
&= \frac{\Gamma_0}{4\pi(b/2)} \frac{\zeta'\,d\zeta'}{(\zeta'-\zeta)\sqrt{1-\zeta'^2}}.
\end{aligned}
\tag{5.62}
$$

Finally, integrating from $-b/2 < z' < b/2$, or $-1 < \zeta' < 1$:

$$
\begin{aligned}
v(z) &= \frac{\Gamma_0}{2\pi b} \int_{-1}^{1} \frac{\eta'\,d\eta'}{(\eta'-\eta)\sqrt{1-\eta'^2}} \\[4pt]
&= \frac{\Gamma_0}{2b} = \text{const.}
\end{aligned}
\tag{5.63}
$$

This above result indicates that, for the elliptical circulation distribution in our current problem, the downwash remains constant along the span. The induced drag D_{ind}, as depicted in Fig. 5.32, results from the varying lift distributed across the wing where:

$$
\begin{aligned}
D_{\text{ind}} &= L \tan\alpha_{\text{ind}} \\[4pt]
&= \rho\Gamma_0 U_\infty \frac{\pi b}{4} \frac{\Gamma_0}{2bU_\infty} \\[4pt]
&= \frac{\pi}{8}\rho\Gamma_0^2,
\end{aligned}
\tag{5.64}
$$

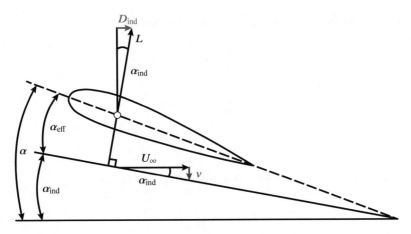

Fig. 5.32 Here we can observe the effect that downwash (v) has on the oncoming flow, where the local wing section *feels* an effective angle of attack (α_{eff}) through the tip-vortex induced flowfield. By turning the oncoming flow through (α_{ind}) the local wing section sees a lower angle of incidence and the lift vector L is tilted backwards, thus producing an additional (inviscid) source of drag, D_{ind}

where $\tan \alpha_{\text{ind}} = v/U_\infty = \Gamma_{\text{o}}/(2bU_\infty)$ and

$$L = \rho U_\infty \Gamma(z)$$

$$= \rho U_\infty \Gamma_{\text{o}} \int_{-b/2}^{b/2} \left(1 - \frac{4z^2}{b^2}\right)^{\frac{1}{2}} dz \tag{5.65}$$

$$= \rho U_\infty \Gamma_{\text{o}} \frac{\pi b}{4}.$$

Note that the induced drag is also a constant, proportional to the center-span circulation Γ_{o}.

Similarly, the induced drag coefficient—note that this drag is inviscid in origin—can be defined as

$$C_{D_{\text{ind}}} = \frac{D_{\text{ind}}}{\frac{1}{2}\rho U_\infty^2 s}$$

$$= \frac{\pi}{4} \frac{\Gamma_{\text{o}}^2}{U_\infty^2 s} \tag{5.66}$$

$$= \frac{\pi}{4} \frac{\Gamma_{\text{o}}^2 AR}{U_\infty^2 b^2},$$

where s is the centerline chord (for a wing with varying chord length), and $AR = b^2/s$ is the wing's aspect ratio.

Finally, the induced drag and induced angle of attack α_{ind} can be related to the lift coefficient:

$$C_{D_{ind}} = \frac{C_L^2}{\pi AR} \quad \text{and} \quad \alpha_{ind} = \frac{C_L}{\pi AR}, \tag{5.67}$$

which provides us with some idea on how high aspect ratios can be beneficial all the while impractical in most other cases, i.e., long appendages are becoming structurally hard to support but of course get in the way with most other aspects of life!

5.7 Helmholtz Vortex Laws and Wake Structure

Returning to our ray, perhaps its elliptical planform, which may ultimately reduce its induced drag, is of no coincidence. Of course, there is more to life than drag reduction, and besides, the ray must *flap* in order to propel itself, which in turn would lead to time-varying circulation across its span.

Further to the above steady description of induced drag for finite wings, in most propulsive systems the appendage oscillates such that the propulsive force (e.g., lift, side force, thrust, etc.), and hence vortex strength, will vary in time and space. In fact, in order to satisfy Kelvin's circulation theorem (Eq. (5.25)), all starting and stopping vortices, in three-dimensional space, will have to balance out. It is for this reason we see simple yet elegant vortex-ring structures in the wake of a jellyfish and ray, as shown in exemplary fashion in Fig. 5.33.

It turns out that, if we could magically *see* the wakes of all swimming and flying animals, their unsteady gaits produce strong vortex rings, sometimes distinct from one another (e.g., jellyfish), sometimes connected in the form of a *ladder*, but always satisfying the *Helmholtz vortex laws* (Helmholtz 1858). Helmholtz identified three basic rules to describe the motion of vortices, and although based on inviscid conditions, they help us make sense of these unique vortex-wake patterns observed in Nature:

1. A vortex filament has constant strength (circulation) at all cross sections;
2. A vortex filament cannot end but must extend boundaries such as walls or form a closed path upon itself, i.e., a vortex ring; and
3. An irrotational fluid parcel will remain irrotational so long as it receives no input from external (rotational) forces.

Of particular interest here is Helmholtz's second law that helps us appreciate the complex reconnections of vortex wakes, as shown exemplarily in Fig. 5.34. Here the (blue) leading-edge vortices (LEVs) are formed on each wing during the downstroke motion and connect to the starting vortex in the wake. Once the downstroke is complete the LEVs will shed and form a closed vortex in the wake not unlike those observed above in Fig. 5.33.

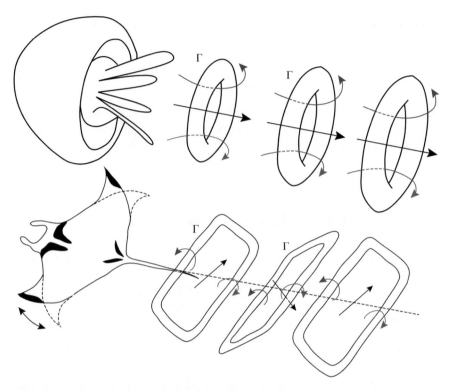

Fig. 5.33 We observe distinct vortex-ring structures formed through oscillatory propulsion behind oscillatory motions of a jellyfish and ray. Distinct vortex-ring patterns are formed in wake providing an indication of time-varying loads on body upstream

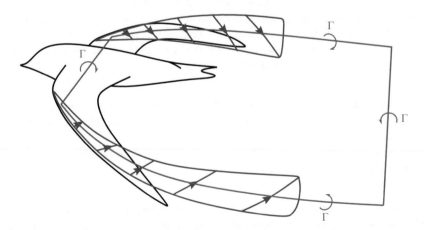

Fig. 5.34 Leading-edge vortices (LEVs) shown in blue are formed during the wing downstroke and connect to the starting vortex in the wake. Once the downstroke motion is complete these vortices shed off and form a closed vortex ring that is left in the wake

5.8 Actuator-Disk Theory

Beyond the autorotation of samaras, one can argue that continuously rotating wings (or blades) do not exist in Nature. However, when considering the ubiquity of hovering in insect flight—and hummingbirds for that matter!—we find some striking parallels, as shown in Fig. 5.35. Although it is tempting to jump right in and try to model the full, three-dimensional vortex structure formed on each wing half-stroke, we will focus here on a simple abstraction of hover, and hover-like propulsion, best described through *actuator-disk theory*, as developed by Froude (1889).

The concept, as depicted in Fig. 5.35, approximates the spinning seed, or flapping insect, as a thin, steady, and *porous* actuator disk by smearing the blades from the hovering body into a uniform disk from which we can easily account for the change in pressure across the rotor plane (Δp_R). To calculate this change (jump) in pressure, we apply Bernoulli's equation (Eq. (5.5)) along the streamline in Fig. 5.36 from position 1 to 2, and then from position 3 to 4, resulting in

$$
\begin{aligned}
1 \to 2: \quad p_1 + \frac{1}{2}\rho u_1^2 &= p_2 + \frac{1}{2}\rho u_2^2 \\
3 \to 4: \quad p_3 + \frac{1}{2}\rho u_3^2 &= p_4 + \frac{1}{2}\rho u_4^2,
\end{aligned}
\tag{5.68}
$$

where in air (or at sufficiently small scales in water) the hydrostatic contribution is effectively negligible ($\rho g \Delta z \ll \frac{1}{2}\rho(\Delta v)^2$). Note that $p_1 = p_4 = p_{\text{atm}}$ and, if we assume the actuator disk is thin, $u_2 = u_3$. It should be noted that when delving deeper into this topic we often relate the local velocity to the upstream velocity

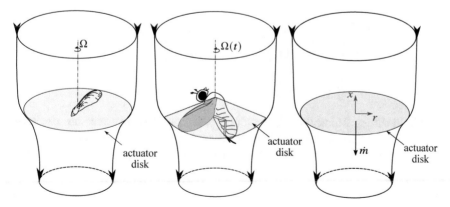

Fig. 5.35 A sudden jump in pressure across the *rotor* plane is abstracted by a uniform actuator disk, as shown on the right-hand side. By replacing the detailed aerodynamics of steadily rotating (Ω)—or even oscillating ($\Omega(t)$)—wings with this simple actuator-disk concept we can focus on the global features of the system so as to develop first insights into the problem in hand

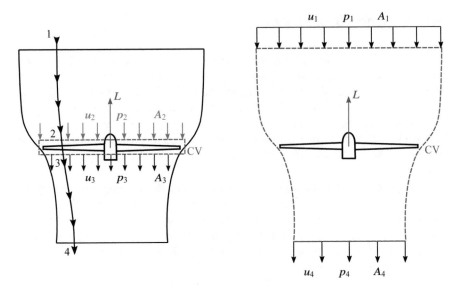

Fig. 5.36 Both infinitely thin (left) and large (right) control volumes used to describe the momentum balance on a generic rotor plane (actuator disk). Note the CV on the right follows the streamtube such that no fluid crosses this circumferential surface

through an *induction factor*, which is a strong function of the amount of thrust (acceleration of the flow). For instance, a slight gust lifting a samara vertically will result in a much weaker pressure rise when compared to a hovering bumblebee, but for the current discussion we will leave this business about induction factors to the side.

Thus, the pressure rise over the actuator disk can be defined as

$$
\begin{aligned}
\Delta p_R &= p_3 - p_2 \\
&= \left[p_{\text{atm}} + \frac{\rho}{2}(u_4^2 - u_3^2) \right] - \left[p_{\text{atm}} + \frac{\rho}{2}(u_1^2 - u_2^2) \right] \\
&= \frac{\rho}{2} \left(u_4^2 - u_1^2 \right).
\end{aligned}
\tag{5.69}
$$

The resulting lifting (hover) force (L), based on the above pressure rise from Eq. (5.69), can then be calculated for the actuator. In Fig. 5.36, we assume and define a thin control volume such that $u_2 = u_3$ since areas $A_2 = A_3 = A_R$ are identical.

We then begin by applying a momentum balance on the small CV in the vertical x-direction:

$$
\sum f_x = \frac{d}{dt} \int_{CV} \rho u \, dV + \int_{CS} \rho u (\mathbf{u} \cdot \mathbf{n}) \, dA.
\tag{5.70}
$$

Tips

In simplifying our momentum balance equation recall that:

- $\frac{d}{dt}\int_{cv}\rho u d\Psi = 0$ for our initial abstracted steady flow;
- $F_x = L$ is the reaction force imparted onto our insect or hummingbird by the fluid; and
- $\int_{CS} p(-\mathbf{n})dA$ represents a pressure force acting *into* the CV (appears on left-hand side of momentum balance).

Subsequently expanding and then simplifying we can write

$$F_x + \int_{cs} p dA = \int_{cs} \rho u(\mathbf{u}\cdot\mathbf{n})dA$$

$$F_x = -(p_2 A_2 - p_3 A_3) + \rho(u_3^2 A_3 - u_2^2 A_2) \qquad (5.71)$$

$$= (p_3 - p_2)A_R$$

$$= \Delta p_R A_R,$$

where $A_R = \pi R^2$ is the surface area of the actuator disk. Substituting in Eq. (5.69), we obtain the lifting force $L = F_x$:

$$L = \frac{\rho}{2}(u_4^2 - u_1^2)\pi R^2. \qquad (5.72)$$

Now, if we zoom out, and consider a large CV that encompasses the streamtube drawn in Fig. 5.36, we can also calculate L based on the momentum flux at the inlet (position 1) and outlet (position 4). Again, we begin with the following force balance for a steady flow:

$$L + \int_{cs} p dA = \int_{CS} \rho u(\mathbf{u}\cdot\mathbf{n})dA$$

$$L = -(p_1 A_1 - p_4 A_4) + \rho(u_4^2 A_4 - u_1^2 A_1) \qquad (5.73)$$

$$= \dot{m}(u_4 - u_1),$$

where $\dot{m}$ is the mass flux through the CV.

Fig. 5.37 Here we observe the smooth increase in velocity throughout the streamtube along with an equivalent jump in pressure across the actuator-disk plane. In practice this pressure rise occurs smoothly between the suction and pressure sides of the hovering wing

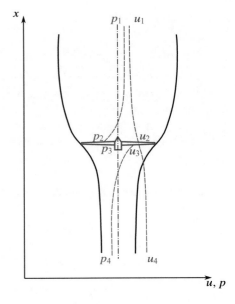

Tips
Note that:

- $p_1 = p_4 = p_{atm}$; and
- $A_1 \neq A_4$, but the pressure force due to uniform p_{atm} cancels out around the large CV.

At this point, we can stop and reflect on how the velocity and pressure fields may be varying through the streamtube, as depicted in Fig. 5.37. Here we can observe that the velocity increases smoothly from inlet to outlet of the streamtube all the while the pressure jumps at the actuator disk plane. Of course in practice, the pressure rise across the hovering wing plane would not form a discontinuity, but rather follow the form of a smooth (continuous) rise equivalent to the pressure difference between suction and pressure sides of the wing.

Next, we can equate Eqs. (5.72) and (5.73) for lift:

$$\frac{\rho}{2}(u_4^2 - u_1^2)A_R = \dot{m}(u_4 - u_1)$$

$$\frac{\rho}{2}(u_4 + u_1)A_R = \dot{m} \qquad (5.74)$$

$$\frac{\rho}{2}(u_4 + u_1)A_R = \rho A_R \bar{u},$$

such that the average velocity $\bar{u}$ at the actuator disk can be defined as

$$\bar{u} = \frac{u_4 + u_1}{2}. \tag{5.75}$$

Using Eqs. (5.72) and (5.75), the resulting actuator-disk velocity ($\bar{u}$) can be defined in terms of the upstream fluid velocity and the lift:

$$\frac{\bar{u}}{u_1} = \frac{1}{2} + \frac{1}{2}\left[1 + \frac{L}{\frac{1}{2}\rho A_R u_1^2}\right]^{1/2}. \tag{5.76}$$

The effort required by a hovering insect can be evaluated in terms of a propulsive efficiency which is the ratio of its propulsive power to the total amount of power imparted on the passing fluid. The propulsive power P_{disk} is simply the product of the lift and disk velocity:

$$P_{\text{disk}} = L\bar{u} = \frac{1}{2}Lu_1\left(1 + \left[1 + \frac{L}{\frac{1}{2}\rho A_R u_1^2}\right]^{1/2}\right). \tag{5.77}$$

The total power expended (P_{tot}) during flight is equal to the change in kinetic energy of the flow as it passes through the actuator disk:

$$P_{\text{tot}} = \dot{m}\left(\frac{u_4^2}{2} - \frac{u_1^2}{2}\right) = \frac{1}{2}\rho\bar{u}A_R\left(u_4^2 - u_1^2\right). \tag{5.78}$$

Finally, the propulsive efficiency η_{disk} is simply the ratio of these two expressions for power:

$$\eta_{\text{disk}} = \frac{P_{\text{disk}}}{P_{\text{tot}}} = \frac{2}{1 + \left(1 + \frac{L}{\frac{1}{2}\rho A_R u_1^2}\right)^{1/2}}. \tag{5.79}$$

Note that η_{disk} is of the exact same form as the efficiency of a jet engine:

$$\eta_{\text{jet}} = \frac{2}{1 + \frac{u_e}{u_o}}, \tag{5.80}$$

where u_e is the exhaust velocity and u_o is the inlet air velocity of the engine.

Since the development of modern aerodynamic theory, at the beginning of the twentieth century, the paradox concerning insect—not just bumblebee!—flight has floated about. The conundrum lies in our observation that conventional *steady* aerodynamic theory, when applied to many disproportionally small insect wings, does not account for sufficient lift to support their respective body weight. For that reason the bumblebee is often mentioned due to the fact that there is an impressive disparity between its wing area and body mass.

It turns out that when examining bumblebee flight—see middle schematic in Fig. 5.35—the actuator disk is in fact quite tiny in comparison to body size. During hover each wing rotates forward and aft sweeping out an approximate disk area. However, on closer observation, it turns out that strong leading-edge vortices (LEVs) are formed on each wing half-stroke, which in turn can double or even triple the local lift coefficient when compared to steady approximations.

In other words, just like when we tread water in a pool, by oscillating our arms back-and-forth, hovering insect wings use strong (periodic) leading-edge separation to overcome gravity in what is often referred to as *drag-based propulsion*. In other words, the bumblebee overcomes gravity in a similar manner to how we use periodic rowing strokes to thrust ourselves forward in a boat!

5.9   Structural Loading on Appendages

Propulsive appendages in the natural world take on a startling broad range of forms, and are derived from completely disparate origins, i.e., contrast the seemingly convergent tail fins of an *Ichthyosaur* with that of a common dolphin. For the current description, let us consider a generic *wing*, from a samara or analogous to a bio-inspired propeller blade, as shown for contrast in Fig. 5.38. Both lifting surfaces are fixed to their respective root sections (origin). This abstracted system is similar to that of a cantilever beam, but now the resultant aerodynamic force, i.e., lift, along with weight, contributes to the distributed loading. In turn, this loading produces a deformation, which in an oscillating system would deform as a function of time. Note that the wing can, as shown previously, be split into a series of two-dimensional slices (or sections). Figure 5.39 depicts one such slice overlaid with the positions of its local elastic axis (x_{EA}), aerodynamic center (x_{AC}), and the forces acting during steady rotation at angular velocity Ω. The lift (C_L) and drag (C_D) produced at this local angle of attack result in normal (C_N) and axial (C_X) forces.

Having established the nomenclature in Fig. 5.39, we can now evaluate the resulting moments. The *flapwise* moment, bending the wing sections laterally along the span, is defined as

$$M_F = \int_0^R \left(\frac{1}{2} \rho U_{\mathrm{eff}}^2 c \right) C_N z \, dz, \qquad (5.81)$$

where U_{eff} represents the local induced (effective) velocity. In a similar fashion, the *edgewise* moment, defined as bending in the chordwise direction, is defined as

$$M_E = \int_0^R \left(\frac{1}{2} \rho U_{\mathrm{eff}}^2 c \right) C_X z \, dz; \quad \text{and} \qquad (5.82)$$

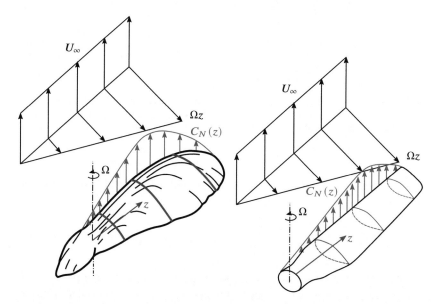

Fig. 5.38 Two representative *wings*, a samara (left) and a bio-inspired propeller blade (right), broken down into two-dimensional slices (or sections) for analysis. We observe relative inflow velocity components (U_∞ and Ωz) in the rotating frame of reference responsible for setting the local aerodynamic forces

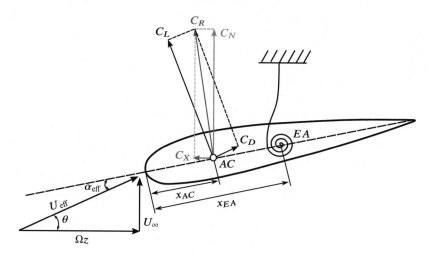

Fig. 5.39 Schematic representation of the forces acting through the aerodynamic center (AC) on a wing (or blade) section, in turn generating moments about the elastic axis (EA)

Fig. 5.40 Reducing 3D deformation to 2D slices. Here a flapwise moment can be reduced to a 2D plunging motion (left), whereas a torsional moment induces pitching (right). In both cases, the instantaneous formation results in a lift where for plunging it is restoring and for pitching typically amplifying

finally the *torsional* moment (or torque), representing the local twisting deformation, is

$$M_T = \int_0^R \left(\frac{1}{2} \rho U_{\text{eff}}^2 c \right) C_R (x_{EA} - x_{AC}) dz. \tag{5.83}$$

It should be noted that the flapwise and edgewise moments (and deformations) are typically stable, but that the torsional moment can be highly unstable. This latter torsional behavior of a wing (or blade) will be treated quantitatively in Sect. 5.10, but first, we can qualitatively characterize these three-dimensional (3D) deformations by reducing them to two-dimensional (2D) motions on a particular section of interest.

Let us begin by approximating 3D (spanwise) bending as a 2D plunging motion, i.e., our airfoil is only free to translate in the normal direction. Figure 5.40 shows an airfoil subjected to an applied vertical disturbance ($\dot{y}$), which produces in turn a response in lift. This lift is in fact a restoring force (opposes the normal translation) since lift is proportional to $|\Delta \dot{h}|$ and $|\dot{h}|$.

Similarly, 3D torsion effects can be reduced to a 2D pitching motion, where the airfoil is free to only rotate about its mid-chord position (in this case the elastic axis EA). The right airfoil shown in Fig. 5.40 is rotated through $\Delta \alpha$ (about the elastic axis). Again, lift is generated in response to the disturbance, but this time it amplifies the torsional moment because both lift and the moment are proportional to $\Delta \alpha$ and $\dot{\alpha}$. The amplification produces a torsional instability (a self-exciting local pitching motion) known as *flutter*. It is worth mentioning though that angular acceleration ($\ddot{\alpha}$) provides a dampening effect in that a rotational added mass opposes the rotational acceleration and helps restore the wing section.

Bending cases are typically stable and can be treated using material strength analysis where we effectively have a stress $\sigma(y)$ in the blade due to bending (akin to a cantilever beam), as illustrated in Fig. 5.41, such that:

$$\sigma(y) = \frac{M_F y}{I_x}, \tag{5.84}$$

where I_x is the second moment of inertia of the wing cross section.

Fig. 5.41 Bending causes a stress in the wing root that is proportional to the flapwise moment

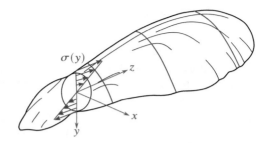

For torsional deformations, which can be unstable, the material stiffness must be considered, i.e., the wing's stiffness must balance the pitching moment. For simplicity's sake, consider a wing that initially has $\alpha = 0°$ and zero twist. The applied torque T is then:

$$T = JG \frac{d\alpha}{dz}, \tag{5.85}$$

where $d\alpha/dz$ is the twist produced by aerodynamic loading, G is the shear modulus, and J is the polar moment of inertia of a section of interest (i.e., the root). Note that the product JG represents the torsional stiffness itself.

Rearranging Eq. (5.85), we can write

$$d\alpha = \frac{T}{JG} dz, \tag{5.86}$$

and then solve for the resultant torsional deflection:

$$\alpha = \int_0^{z_1} \frac{T}{JG} dz, \tag{5.87}$$

where z_1 is the position along the span where equilibrium is calculated. The resultant twist along the wing due to torsional loading can be calculated such that:

$$\begin{aligned}
\alpha &= \int_0^{z_1} \frac{1}{JG} \int_{z_1}^{R} T \, dz \, dz \\
&= \int_0^{z_1} \int_{z_1}^{R} \left(\frac{1}{2} \rho U_\infty^2 c \right) C_L (x_{EA} - x_{AC}) \, dz \, dz,
\end{aligned} \tag{5.88}$$

where, for steady flow and low α, $C_L = 2\pi\alpha$ (based on potential-flow theory).

5.9.1   *Exercise for Albatross Take-Off Process*

Let us consider the extreme root loading of an albatross wing during its remarkable
take-off process, as shown in Fig. 5.42a. The albatross, with its awkwardly long
wings that cannot flap at ground level, begins running and picking up speed before
taking a final leap of faith. At the moment of take-off, the albatross is shown flying
upward with an angle of 30° with respect to the ground, and with a take-off speed
of $U_\infty = 15$ m/s. For sake of abstraction, the wing is assumed to be rigid in this

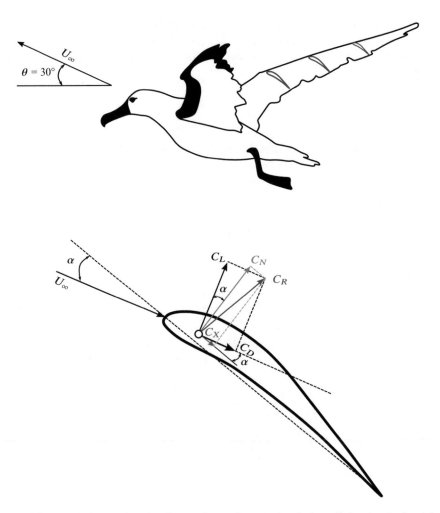

Fig. 5.42 (top) Moment when the albatross leaves the ground and takes off showing the inertial
path relative to the ground (horizontal plane). (bottom) Here, we see the relative inflow and the
associated resultant, normal, and axial forces (C_R, C_N, and C_X) at each wing section (we are
assuming no twist along the span)

example and is modeled with an elliptical wing planform with half-span $b/2 = 1$ m and a root chord of $c_r = 0.2$ m. The spanwise distribution of the chord length is given by

$$c(z) = c_r \sqrt{1 - 4\left(\frac{z}{b}\right)^2}. \tag{5.89}$$

The steady aerodynamic forces acting on a wing cross section are shown in Fig. 5.42b during the moment of lift off. The angle of attack of the wing is 40° with respect to the ground, and thus the wing section experiences an angle of attack of $\alpha = 10°$. From potential-flow theory, the sectional lift can be approximated by $C_L = 2\pi\alpha$ for steady, attached-flow conditions. Furthermore, we will assume $C_D = 0.04$ for this example. Furthermore, we will ignore root or tip effects on these high aspect ratio wings to keep the analysis simple.

To obtain the structural moments at the wing's root, which is expected to be the location with the highest loading across the wing span, we first need to calculate the resultant, normal, and axial forces (C_R, C_N, and C_X) at each wing section:

$$C_R = \sqrt{C_L^2 + C_D^2}$$
$$= 1.095, \tag{5.90}$$

$$C_N = C_L \cos(\alpha) - C_D \sin(\alpha)$$
$$= 1.071, \tag{5.91}$$

and

$$C_X = C_L \sin(\alpha) + C_D \cos(\alpha)$$
$$= 0.229. \tag{5.92}$$

Therefore, the *flapwise* moment at the wing's root can be obtained using Eq. (5.81):

$$M_F = \int_0^{b/2} \left(\frac{1}{2}\rho U_\infty^2 c\right) C_N z \, dz$$
$$= \frac{1}{2}\rho U_\infty^2 C_N \int_0^{b/2} c z \, dz$$
$$= \frac{1}{2}\rho U_\infty^2 C_N \int_0^{b/2} c_r \sqrt{\left[1 - 4\left(\frac{z}{b}\right)^2\right]} z \, dz \tag{5.93}$$
$$= \frac{c_r b^2}{24}\rho U_\infty^2 C_N$$
$$= 9.839 \text{ Nm}.$$

Similarly, the *edgewise* moment at the root section is calculated as per Eq. (5.82):

$$
\begin{aligned}
M_E &= \int_0^{b/2} \left(\frac{1}{2} \rho U_\infty^2 c \right) C_X z \, dz \\
&= \frac{1}{2} \rho U_\infty^2 C_X \int_0^{b/2} cz \, dz \\
&= \frac{1}{2} \rho U_\infty^2 C_X \int_0^{b/2} c_r \sqrt{\left[1 - 4 \left(\frac{z}{b} \right)^2 \right]} z \, dz \\
&= \frac{c_r b^2}{24} \rho U_\infty^2 C_X \\
&= 2.104 \text{ Nm.}
\end{aligned}
\tag{5.94}
$$

Finally, the *torsional* moment (Eq. (5.83)) at the root, assuming $x_{AC} = 0.25c$ and $x_{EA} = 0.4c$, can be calculated as

$$
\begin{aligned}
M_T &= \int_0^{b/2} \left(\frac{1}{2} \rho U_\infty^2 c \right) C_R (x_{EA} - x_{AC}) \, dz \\
&= \frac{1}{2} \rho U_\infty^2 C_R (x_{EA} - x_{AC}) \int_0^{b/2} c \, dz \\
&= \frac{1}{2} \rho U_\infty^2 C_R (x_{EA} - x_{AC}) \int_0^{b/2} c_r \sqrt{\left[1 - 4 \left(\frac{z}{b} \right)^2 \right]} \, dz \\
&= \frac{\pi c_r b}{16} \rho U_\infty^2 C_R (x_{EA} - x_{AC}) \\
&= 3.554 \text{ Nm.}
\end{aligned}
\tag{5.95}
$$

Take-off is often a very precarious moment, particularly for ground-dwelling birds that cannot take advantage of perches. The albatross's wings are so long that to make life worse, it cannot properly flap without hitting its wings on the ground during this initial phase. Nevertheless, we are able to get a first-order estimate of the structural loadings in the shoulder region with these simple approximations. Of course once airborne, the loads might be even higher when one accounts for the dynamics of flapping as well as sudden bursts of gusts at even higher relative wind speeds.

5.10   Fluid–Structure Interactions on Appendages

As we have seen in Sect. 4.6, fluid–structure interaction is the study of coupled structural and aero/hydrodynamic loading. Furthermore, despite the necessary abstractions discussed above, nearly all forms of propulsion in Nature are indeed flexible, e.g., feather deflections, tail–fin deformations, etc. Of course, more complex behavior is observed in a coupled system than when examining either discipline individually, as shown schematically in Fig. 5.43.

The feedback loop shown in Fig. 5.43 has the potential to produce highly destructive forces but, conversely, can also provide enormous benefits in terms of efficiency and/or maneuverability, often left unharnessed in engineered systems. Furthermore, we can envision a broad range of challenges faced in Nature, such as how to control stable gliding with feathers or membranes, particularly when operating in gusty environments. On the other end of the spectrum, many unsteady propulsive cases represent active forcing from the body where, for instance, muscle input helps stiffen the structure during the motion.

Let us start simple and consider the abstracted case of a two-dimensional profile with only one pitching degree of freedom, as shown in Fig. 5.44.

If we first assume steady-state and attached-flow conditions, such that $C_L = 2\pi\alpha$, then structurally, the system can be modeled as a classic spring-mass-damper:

$$J\ddot{\alpha} + c\dot{\alpha} + k\alpha = M, \qquad (5.96)$$

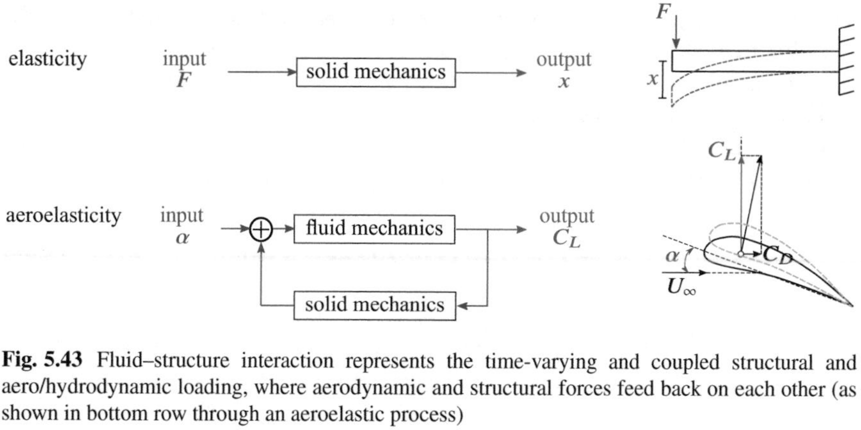

Fig. 5.43 Fluid–structure interaction represents the time-varying and coupled structural and aero/hydrodynamic loading, where aerodynamic and structural forces feed back on each other (as shown in bottom row through an aeroelastic process)

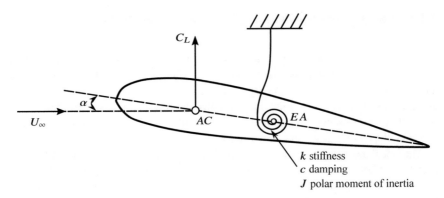

Fig. 5.44 Abstraction of a profile with a single pitching degree of freedom undergoing an aeroelastic response, i.e., experiencing fluid–structure interaction

where J is the (polar) moment of inertia, c represents damping, and k the stiffness. Of course, the relative contribution of damping and inertia (mass) is a strong function of both material (e.g., tissue) properties and the surrounding fluid medium. The *mass ratio*, defined as

$$m^* = \frac{m}{\rho \mathcal{V}}, \tag{5.97}$$

represents the relative density of the body to its surroundings, which can vary by many orders of magnitude when contrasting the movement of feathers in air to say the dynamics of tissue in water, as just one crude example.

While to start, steady lift models are used here, we should be conscious that more complete system modeling becomes more complex and requires the following:

$$J\ddot{\alpha} + c\dot{\alpha} + k\alpha = J_{\text{aero}}\ddot{\alpha} + c_{\text{aero}}\dot{\alpha} + k_{\text{aero}}\alpha, \tag{5.98}$$

where J_{aero}, c_{aero}, and k_{aero} account for additional attenuation and phase-shift behavior. In particular, added mass can play a very critical aspect of these problems and can dominate over the actual system mass, particularly when considering problems in liquids such as water.

For the moment though, let us focus on a simple case. We can determine the steady-state pitching moment from our instantaneous lift force as

$$M = \left(\frac{1}{2}\rho U_\infty^2 c\right)(2\pi\alpha)(x_{EA} - x_{AC}), \tag{5.99}$$

where the moment is proportional to the angle of attack (α) and therefore behaves like a negative stiffness:

$$J\ddot{\alpha} + c\dot{\alpha} + k\alpha = k_{\text{aero}}\alpha = \left(\frac{1}{2}\rho U_\infty^2 c\right)(2\pi)(x_{EA} - x_{AC})\alpha \tag{5.100}$$

or

$$J\ddot{\alpha} + c\dot{\alpha} + \left[k - \left(\frac{1}{2}\rho U_\infty^2 c\right)(2\pi)(x_{EA} - x_{AC})\right]\alpha = 0, \tag{5.101}$$

$$J\ddot{\alpha} + c\dot{\alpha} + k_{\text{tot}}\alpha = 0.$$

We see that when x_{EA} is located in front of x_{AC} (i.e., $x_{EA} - x_{AC} < 0$), then overall the system will become stiffer ($k_{tot} \geqslant k$). Although $x_{AC} \approx 0.25c$ for traditional cross sections, this position can be highly variable when considering aerodynamic sections such as those on thin bird and bat wings. Therefore, the system can easily become much less stiff ($k_{\text{tot}} \leqslant k$), which, in combination with aerodynamic forces that increase with U_∞, results in a decrease of the overall total stiffness of the coupled system (combination of structural and fluid forces).

Once the system stiffness reaches zero, deflections can grow in a process known as *divergence*:

$$k_{\text{tot}}(U_{\text{div}}) = 0. \tag{5.102}$$

Thankfully, U_{div} can easily be calculated to test system stability by setting total stiffness to zero (from Eq. 5.101):

$$k - \left(\frac{1}{2}\rho U_{\text{div}}^2 c\right)(2\pi)(x_{EA} - x_{AC}) = 0$$

$$\frac{1}{2}\rho U_{\text{div}}^2 c = \frac{k}{(2\pi)(x_{EA} - x_{AC})} \tag{5.103}$$

$$U_{\text{div}} = \sqrt{\frac{k}{(\frac{1}{2}\rho c)(2\pi)(x_{EA} - x_{AC})}}.$$

Recall more generally that the natural frequency of a system varies with stiffness:

$$\omega_n = \sqrt{\frac{k_{\text{tot}}}{J}}, \tag{5.104}$$

where ω_n is the response frequency to a disturbance and tends toward zero near $U = U_{\text{div}}$.

As a final step, if we consider the sectional pitching moment (M) as a function of instantaneous angle of attack (α), fluid energy can either be injected into the

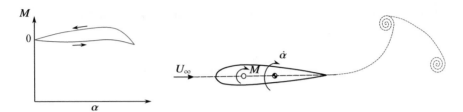

Fig. 5.45 Here, as an example, we observe *hysteresis* in the moment-alpha $(M - \alpha)$ curve over one oscillation period for a dynamic-pitching $(\dot{\alpha})$. Since the loop is counterclockwise it depicts energy transferred from airfoil to fluid such that the power can be estimated through integration of Eq. (5.105)

surrounding fluid from the structure (i.e., propulsive power) or returned to the structure from the fluid, as shown by

$$P = \frac{1}{T} \int_0^T M(t)\alpha(t)dt, \qquad (5.105)$$

where P represents the power input to the structure and T is the period of oscillation. In other words we can characterize fluid energy transferred into the structure (airfoil) through a clockwise loop, or conversely, energy removed from the structure (airfoil) to the fluid through a counterclockwise loop. This effect is qualitatively shown in Fig. 5.45 for this simple dynamic-pitching case. This general concept can easily be extended to more complex motions where, for instance, the local wing (or appendage) section has a plunging degree of freedom.

References

Babinsky, H. (2003). How do wings work? *Physics Education, 38*(6), 497–503.

Betz, A. (1912). Ein Beitrag zur Erklärung des Segelfluges betz. *Zeitschrift für Flugtechnik und Motorluftschiffahrt, 3*, 269–272.

Biot, J.-B., & Savart, N.-P.-A. (1820). Note sur le magnétisme de la pile de Volta. *Annales de Chimie et de Physique, 15*, 222–223.

Birnbaum, W. (1924a). Das ebene Problem des schlagenden Flügels. *Zeitschrift für Angewandte Mathematik und Mechanik, 4*(4), 277–292.

Birnbaum, W. (1924b). Der Schlagflügelpropeller und die kleinen Schwingungen elastisch befestigter Tragflügel. *Zeitschrift für Flugtechnik und Motorluftschiffahrt, 15*(11), 128–134.

Dececchi, T. A., Larsson, H. C., & Habib, M. B. (2016). The wings before the bird: An evaluation of flapping-based locomotory hypotheses in bird antecedents. *PeerJ, 4*, e2159.

Fernando, J. N., & Rival, D. E. (2017). On the dynamics of perching manoeuvres with low-aspect-ratio planforms. *Bioinspiration & Biomimetics, 12*(4), 046007.

Froude, R. E. (1889). On the part played in propulsion by differences of fluid pressure. *Transactions of the Institution of Naval Architects, 30*, 390–405.

Helmholtz, H. V. (1858). Über Integrale der hydrodynamischen Gleichungen, welche den Wirbelbewegungen entsprechen. *Journal für die reine und angewandte Mathematik, 55*, 25–55.

Joukovsky, N. E. (1906). On annexed [bounded] vortices. *Trudy Otdeleniya Fizicheskikh Nauk Obshchestva Lubitelei Estestvoznaniya, 13*(2), 12–25. [in Russian].

Katzmayr, R. (1922). Effect of periodic changes of angle of attack on behavior of airfoils. *NACA report No. 147 (translated from Zeitschrift für Flugtechnik und Motorluftschiffahrt)* (pp. 95–101).

Knoller, R. (1909). Die Gesetze des Luftwiderstandes. *Flug- und Motortechnik (Wien), 3*(21), 1–7.

Kutta, W. M. (1902). Auftriebskräfte in strömenden Flüssigkeiten Ill. *Aeronaut Mitt, 6*, 133–135.

Lilienthal, O. (1889). *Birdflight as the basis of aviation*.

Milne-Thomson, L. M. (1968). *Theoretical hydrodynamics*. Macmillan Education UK.

Polet, D. T., & Rival, D. E. (2015). Rapid area change in pitch-up manoeuvres of small perching birds. *Bioinspiration & Biomimetics, 10*(6), 066004.

Prandtl, L. (1918). *Tragflügeltheorie*. Königliche Gesellschaft der Wissenchaften zu Göttingen.

Thomson, W. (1868). VI.—on vortex motion. *Transactions of the Royal Society of Edinburgh, 25*(1), 217–260.

von Kármán, T., & Sears, W. R. (1938). Airfoil theory for non-uniform motion. *Journal of the Aeronautical Sciences, 5*(10), 379–390.

Wagner, H. (1925). Über die Entstehung des dynamischen Auftriebes von Tragflügeln. *Zeitschrift für Angewandte Mathematik und Mechanik, 5*(1), 17–35.

Weymouth, G. D., & Triantafyllou, M. S. (2013). Ultra-fast escape of a deformable jet-propelled body. *Journal of Fluid Mechanics, 721*, 367–385.

Printed in the United States
by Baker & Taylor Publisher Services